You Are Here

You Are Here

Exposing the Vital Link

Between What We Do and

What That Does to Our Planet

Thomas M. Kostigen

HarperOne
An Imprint of HarperCollins*Publishers*

HarperOne

HarperCollins books may be purchased for educational, business, or sales promotional use. For information please write: Special Markets Department, HarperCollins Publishers, 10 East 53rd Street, New York, NY 10022.

HarperCollins Web site: http://www.harpercollins.com
HarperCollins®, 🏭®, and HarperOne™ are
trademarks of HarperCollins Publishers

FIRST HARPERCOLLINS PAPERBACK EDITION PUBLISHED IN 2010

Designed by Level C

Library of Congress Cataloging-in-Publication Data
Kostigen, Thomas.
You are here : the surprising link between what we do and what that
does to the planet / by Thomas M. Kostigen. — 1st ed.
p. cm.
Includes bibliographical references and index.
ISBN 978–0–06–158037–6
1. Nature—Effect of human beings on.   2. Environmental
protection—Citizen participation.   3. Environmentalism.   I. Title.
GF75.K674   2008
363.7—dc22   2008010598

Mixed Sources

Product group from well-managed
forests, controlled sources and
recycled wood or fiber
www.fsc.org Cert no. SCS-COC-00648
© 1996 Forest Stewardship Council

10 11 12 13 14   RRD(H)   10 9 8 7 6 5 4 3 2 1

In memory of my father,
Walter S. Kostigen

Contents

Foreword   ix

Introduction   1

ONE   Losing Our Past
       Jerusalem, Israel   5

TWO   Our Future
       Mumbai, India   21

THREE We Are Not Alone
       Borneo, Southeast Asia   41

FOUR   What Are We Doing?
       Linfen City, China   61

FIVE   Distant Consequences
       Shishmaref Village, Alaska   81

SIX   Nature's Oxygen Factory
       The Amazon Jungle   103

SEVEN The Reality of Our Actions
       The Fresh Kills Landfill, New York   123

EIGHT   Where the Currents Take Our Trash
        The Eastern Garbage Patch,
        Pacific Ocean   143

NINE   The Greatest Problem No One
        Has Heard About
        The Great Lakes, Duluth, Minnesota   167

TEN   Where to Spread Your Wings
        Home: Santa Monica, California   191

        Afterword   207

        Acknowledgments   211

        Resources   213

        Index   245

        About the Author   257

Foreword

Six degrees of Kevin Bacon. When I first heard about the game I was horrified. I looked at it as a joke at my expense: Can you believe this mediocre actor can be connected to all the greats in six degrees or less? Kevin Bacon to Lawrence Olivier? I thought for sure it would go away, fade into a distant memory like pet rocks and the preppy handbook. But it had such a long hang time. I turned around, ten years had passed, and more and more people were mentioning it to me—at parties, on the subway. "Hey Kev, does this mean I'm one degree?" Amazing, really. After all, it's just a concept. It's not a thing that you can buy or sell (I've tried). And if you take me out of the equation, it's a beautiful notion, the idea that we are all connected: That what we do, our actions, affect our friends and neighbors down the block and on the other side of the world; that we must be responsible for each other and for this planet we are all riding on.

This was what inspired me to launch www.sixdegrees.org last year. And I think there is a similar quality of inspiration that makes this book so important: In it are the ripple effects that I have been talking about. These are the amazing wake up calls we need to understand and become aware of so we can embrace change – for a better world and planet that we all just happen to be living on together.

It's eye opening to follow how kids in Africa are affected when your cell phone is tossed; how polar bears in the Arctic are affected when your light switch is turned on; or how the toothpaste you use affects things way out in the middle of the jungle halfway around the world—among the many other connections you'll find in these pages. *You Are Here* is an apt title.

Look, if we are given the right type of information, we'll do the right thing. I firmly believe that. But we need to understand the connections to what we do. We also need to understand that these connections lead back to us.

For example, mindlessly leave a plastic bag or bottle behind when you're outside. It will likely end up being blown into a river or stream that likely will make its way to the ocean where it likely will break apart into smaller pieces, which are likely eaten by fish, which are likely caught and shipped to a market, where the fish gets bought, taken home, and eaten by us. And that's just the fate of one action, in six degrees or less.

When we're more mindful, we're minding the health of ourselves and the planet too. When we understand our place in the chain of events that lead to better place, we'll take that path every time. Awareness is the key.

So if we could remember that every one of us on this planet is connected through six degrees of separation, that we all climbed out of the same swamp, maybe we wouldn't be so quick to rush to war or to turn our backs on our brothers and sisters in need, or do things that even many times unbeknownst to us are bad for the planet.

In this book Tom shows us that our six degrees of separation to the planet is often just one, and that we have to begin to care more for it before that six degrees takes on a different meaning: temperature rise and global warming.

Right now, as the title says, you are here, and with these words you are connected to me. The goal of this book and www.sixdegrees.org is for you to understand where you are in the scheme of the world by what you do and don't do. Of course, we hope with this understanding you'll make the right choices and decisions that will be better for us all, planet included.

Let's actually start to think of it more as one degree of separation—cause and effect. It may just spark a whole new cultural movement.

Kevin Bacon

Introduction

After the runaway success of Al Gore's *An Inconvenient Truth* several years ago, it became clear that people were bent on finding ways to become more environmentally friendly in every aspect of their lives. *An Inconvenient Truth* raised important issues and left people wondering what they could do to help stop global warming.

To address this need, I wrote *The Green Book: The Everyday Guide to Saving the Planet One Simple Step at a Time,* a book that provides solutions—more than four hundred—that people can easily adopt into their everyday lives. It put environmental issues into accessible language and appeared at a time when the green frenzy was just beginning: I traveled the country speaking about the tips and advice in that book. I answered hundreds of e-mails, fielded question after question, and watched as the green marketing machine took off. Eco-friendly products began appearing on store shelves. Hybrid car sales began to sizzle. Lots of "green" homes hit the market. Green rock concerts were staged. Green television shows aired. One Web site even offered green sex tips. Clearly the green movement had entered the mainstream.

But something was getting lost in the rush to market. The issues were getting diluted. The reasoning and rationale for becoming environmentally friendly were becoming commercialized to the extent that people weren't buying "green" anymore, they were being sold it. There was and continues to be a general lack of understanding about why what we do matters.

I began to ask a very simple question after every green solution I heard: "Who cares?" And by that I meant who beyond just me are these problems and solutions affecting?

I couldn't put many faces or many images to the answer to that question; data, sure, but I was at a loss to connect actual people, places, and things.

This book puts images to actions. It is an effort to move beyond the noise, beyond the unbearable weight of the problems.

You will get outside the house, your comfortable and known world, and be taken to places you may have only heard about. You will become an environmental voyeur. You will see what's affecting the world and at the same time become empowered to change its course.

This is not green-lite. It's an adventure story structured to take you on a journey of understanding.

You will be transported into the thick of the most environmentally tenuous places on the planet. The hope is that doing so will create a sense of appreciation for the world and for how each of us, individually, can effect change for the future. Disaster will occur only if we ignore the Earth's problems and stand by and do nothing and leave the problems up to others to fix. It's impossible to take ourselves as individuals out of the equation that will cure the environmental ills of the world. We are here; we are contributing to these ills in surprising ways, ways in which many of us are unaware. We may reduce, reuse, recycle. So we save a tree. We use less gas. We conserve power. What effect do those actions really have on the world? So much of this information is in a vacuum—it is lacking a necessary context. We have been told, not shown, which issues matter and why.

Read on and you will be taken to the frontlines of the environmental battlegrounds. This isn't about conjecture or the future. In these pages, we travel to distant and exotic locations to make clear the price of our current actions.

You'll see just how far-reaching the effects of our actions are and where they end up—in the middle of the ocean, deep in the jungle.

Simply put, we can all continue to enjoy the world's natural resources without great sacrifice. We need to understand, however, which issues we should be focused on and why. For the green movement to continue and go on beyond a fad, we need to have a good grip on the matter of caring. Caring about the effects of our actions is what will make all these green things we do sustainable.

I am not an all-sum environmentalist who believes wearing hemp and eating only "dead" fruits and vegetables will save the world. This kind of extremism is off-putting and does not advance the issues at hand. I am out there in the world stumbling, fumbling, and mumbling around much like you. I am not one to preach. But within these pages you will find one real plea—that's right: Care.

Everything else, I believe, will follow. Of course I try to provide examples of what people should care about and what will be in our and the planet's best interest. You'll be the judge of whether any of that sticks.

There is too much passing the buck these days—that businesses must change first, that the government must set new policies in order for anything to really make a difference. It is easy to write off any real responsibility. Regardless of who does what, we as individuals have to let businesses and governments know that we indeed have a say in our future. To fully understand the issues, we have to be informed about the world in which we live and what environmental issues are trying it most.

So buckle up, settle in, and take note of the ride that follows. We are off the road and exposing the most provocative environmental issues of our time. Trust me; you will care about what you read.

ONE

Losing Our Past

Jerusalem, Israel

The light blinds me after I switch it on. It's 4:00 a.m. and still dark outside. I am awake to ready for my trip to Jerusalem.

Jerusalem may seem like an odd place to begin the journey on which I am embarking. We don't associate it with environmental crisis as we do the melting ice caps, or the dwindling forests. But Jerusalem is indeed tied to the environment in two essential ways.

First, there is a religious connection. Jerusalem is the epicenter of faith for nearly half the world's population. For Christians, Jews, and Muslims, Jerusalem is the holiest place on Earth. It physically and spiritually represents the roots of their belief system. And these beliefs are all based on loving thy neighbor, on community, and on being responsible stewards of the Earth—in other words, caring. Caring for the environment is the first step.

Second, historic monuments are omnipresent in Jerusalem, and those monuments are decaying at a faster rate than at any time in history because of climate change. Indeed, you could say that we are losing not only our future to global warming but our past as well.

‒   ‒   ‒

IN THE BASEMENT of the Ecce Homo Convent of the Sisters of Sion, I jump off some scaffolding and make my way around a dark archway. A few lightbulbs hang from the ceiling fifty or so feet above my head. The mucky silt on the ground exposes my footprints.

Omar, a local tour guide I've convinced with 100 shekels to take me down to this place, stops on a raft of rocks a few feet away from me where big boulders have fallen or been set aside, probably when this maze of pools was built in the second century. We're in the belly of the ancient water system that was used to supply the Temple Mount and the Old City.

Underground arches like the one I am next to rise to where they meet the road above: the Via Dolorosa, where Christ is believed to have walked strapped to a cross. The farthest arch from me is walled off. Just beyond that arch is the end of the Western Wall, a place significant in Jewish religious history as a physical reminder of the destruction of the Second Holy Temple. This point is also the westernmost side of the fortress that holds the Dome of the Rock, Islam's oldest religious monument in existence. We are within a few feet of where the three religions geographically meet.

Drips make their way to puddles and echo throughout the cavern. Omar points: "That's it, there."

Directly in front of me now is the walled archway. Omar is pointing to the top stone in its cap. The silt floor looks like quicksand, or wet cement. "Is it safe to walk across?" I want to get a closer look.

"No," Omar says, ". . . maybe."

One delicate footstep at a time, I make my way over . . .

‒   ‒   ‒

STANDING NOW FAR above that basement, I am at the uppermost corner of the Western Wall. From here I can see all the way across the grounds to the gold Dome of the Rock. This massive mosque sits mightily atop the Temple Mount. I am standing in the middle of an elementary schoolyard. Children play and taunt and scream. It must be recess.

"Hey mister," they yell. "What you do?"

What I am doing is piecing together the stones that comprise the connection to a common God. Specifically, I am looking for just one, the one where all three religions physically meet. I want to see if it too is corroding.

As the Earth's temperature rises it accelerates the effects of pollution. Warmer air concentrates harmful gases and makes them more powerful. It's similar to bleach that has not been diluted with water. The result is that historical sites and ruins begin to decay faster.

This is important because we need to preserve the past. We need to know where we've been to understand where we are going. While archaeology, relics, and ruins may not be your thing, these remnants of our past are the only remaining physical reminders of what once was—our history, our culture, and our story as a people. Here in Jerusalem, this connection to the past is more powerful than anywhere else on the planet.

From an environmental perspective, the past is the best measure of things to come. It always has been. Certain weather cycles exist as they have for centuries. Certain wind patterns occur at the same time each year. On a very basic level, the change of seasons informs us of temperature changes because we've experienced those seasons before. Think about the first time someone lived through winter. I bet the next year they buttoned up.

So a little thing like a stone crumbling in the Middle East may not seem to be much of a big deal, but the implications for the planet are widespread.

"Come look, I'll show you," says Haroot Hammad, age thirty. He grew up in this small patch of Jerusalem's Old City. Haroot runs an antique shop across from the convent of Sion and next to an Israeli guard station. In the rear of his shop he pulls back a carpet hanging on the wall. A massive stone, rough and jagged, is revealed. "This is the original wall to the Via Dolorosa," he says.

Over time, walls, facades, and even street stones erode and are rebuilt. There are lots of cover-ups in the Old City. The Old City looks like one of those ancient villages you see on television or in the movies. It is walled, and there are turrets. Little alleys are called streets, and markets and bazaars are held on many of the narrow passageways. The rest of Jerusalem has sprung up around its walls. To get into the Old City, you have to enter through one of its eight gates.

The Old City is like a living museum for Christians, Jews, and Muslims because they all make claim to various parts of it as testament to their beginnings. There are a lot of questions about where sacred things really are versus where they are supposed to be. And there is supreme controversy over property rights.

Haroot tells me to go outside and look at the stones on the wall and then come back inside and look at those in the back of his store. When I compare, the stones inside are much more of a brown color and are bigger than those outside. The stones outside are tarnished, blackened with soot. The stones inside Haroot's shop look better preserved. They are like the stones I found underground the night before when I took an organized tunnel tour of the Western Wall excavation.

The underground tour of the Western Wall excavation began at what is called the "secret passage" at the far corner of its plaza.

I went through the entrance and made my way past water tunnels, quarries, and pools. I touched the stone entrance to the site of the Holy of Holies, the most sacred site in Judaism, which is the inner sanctuary of the Temple, and ducked through a mine shaft–like tunnel to the wall's end.

It was dark when I eventually emerged from those tunnels through the guard station and onto the Via Dolorosa. It began to rain lightly, and the limestone that paves the Via Dolorosa was slick. Haroot's shop was closed—all of the shops on the street were shuttered. The elementary school above was quiet. It was eerie and creepy as I walked down the narrow street.

There were few people about in this quarter of town. Some teenagers roamed and a man straight out of central casting for a shepherd rambled along. He wore a ghutra and igal—a white headdress scarf fastened by a black cord.

It grew darker at the corner of the Via Dolorosa at the Third Station of the Cross, where I had to turn left onto another small street toward the Wall Plaza. The Third Station of the Cross is where Jesus is said to have fallen for the first time. Overhangs recede into darkness. Alleys call to desolation and dread.

When I stepped onto the Western Wall Plaza, there was a roar of prayer and chatter. Floodlights surrounded the Kotel, the traditional Jewish name for the wall. Hasidim in their black hats and frocks, soldiers in their uniforms, and tourists donning cardboard yarmulkes handed out at the Wall's entrance converged. It was eleven o'clock at night, yet hundreds of people crowded the place. Idling buses and taxis waited for passengers.

Then I walked right smack into the problem.

On the Western Wall Plaza, there are newly constructed wooden bridges and cordoned off areas of construction where stone walkways and staircases stood the last time I was here several years ago. The structures have fallen down. Now there are scaffolding and tarps about. Instead of mounting stone steps and railings that people for centuries made pilgrimage upon, people file up wooden ramps made of plywood.

"What happened?" I ask my friend Matt Beynon Rees, a crime novelist. Matt, a Welshman, lives in Jerusalem and is somewhat of an aficionado of local historical sites. "It rained," he says, deadpan.

Over time, structures weaken. That is to be expected. Yet we aren't paying enough attention to the impacts of weather and climate change—even after Hurricane Katrina and the tsunami. A simple drizzle of rain can seep through cracks, corrode mortar, and let crumble whatever structure is being held together, as it has here at the Western Wall Plaza. How much of it will be left the next time I return?

We ignore reality based on some belief that everything is supposed to be all right. The rain will stop. The weather will normalize. Unfortunately that isn't the case anymore. The climate is changing.

The planet is becoming overcrowded, and in many places cannot provide people with the resources they need to survive and flourish. All the while, we pound it with our feet, rake it with our appetites, and soil it with our waste. This results in less space to grow food, and fewer places to harbor clean water and fresh air. It would now take the resources of five planet Earths to support the current world's population at US standards of living. That's why the world is strained, with nearly a billion people going without a meal each day, and more than a billion lacking access to safe drinking water.

Moreover, we are not being nearly as efficient with the resources we are given. The average-sized house in a temperate climate can fully provide enough water to its inhabitants purely from the rainwater that falls on its roof. The sun provides in one second enough energy to power the entire US population for nine million years, yet we harness less than 1 percent of its energy. The third most common refuse at dumpsites is food.

It's easy to see how things begin to fall apart when we don't care.

The air I'm breathing as I stare at the Western Wall has 37 percent more pollution in it than the air the craftsmen breathed when the ancient structures of the Old City were built. Much of that pollution is carbon dioxide. It's produced when we use certain types of energy, such as oil or coal. Just a little carbon dioxide holds a significant amount of heat. It attracts the sun's rays and keeps heat stored. That's why carbon dioxide is considered a greenhouse gas. Along with methane, and different forms of nitrogen, it adds to the Earth's temperature rise.

There is more pollution now from those gases than in ancient history because there are more people on the planet and we use more energy for our needs: electricity, heating, cooling, manufacturing, transportation—all of the power-related things associated with modern-day living. Energy derived from fossil fuels—gas, or coal, or oil—is an especially big contributor to the concentration of greenhouse gases in the air.

Greenhouse gases are mostly water vapor, carbon dioxide, methane, nitrous oxide, and ground-level ozone. When they get released they dissipate into the air and travel along wind currents. Some of these gases are emitted from man-made activities, but others are produced naturally as part of the Earth's ecosystem.

Those gases and the extra heat they keep trapped in the atmosphere are linked to a number of climate-related outcomes: increased rain frequency and intensity in some places, severe drought in others; flooding; stronger storm surges; coastal erosion; warmer ocean temperatures; rises in sea levels; and extreme weather events.

Besides weather events, there are other ways greenhouse gases impact ancient structures like the one I am looking at.

The Western Wall is made of limestone. Limestone has been widely used throughout history to build things because it's common and durable. But the kryptonite, if you will, to limestone's strength is acidity.

When coal is burned to produce the steam that turns turbines and makes electricity, sulfur dioxide is a byproduct. Acid rain, acid snow, and acid fog occur when sulfur dioxide is released into the air.

So even though my friend Matt was being glib when he said rain was responsible for the crumbling staircases and platforms in the Western Wall Plaza, he was right.

As a growing country in the Middle East, Israel gets wafts of air pollution from oil refineries. It's a desert country so it needs air conditioning and cooling. The energy that drives those things comes from coal plants, and coal plants send sulfur into the air like there's no tomorrow, excuse the pun. When it rains, the sulfur comes pouring back down to earth and wears down its structures. In Jerusalem, most notably, those limestone structures just happen to represent religious monuments and our religious history.

Much of the concern surrounding acid rain has to do with its impact on lakes, streams, wildlife, trees, and soils. However, acids also accelerate the corrosion of stone and metal by reacting with the compounds in the material. The rate of decay depends on the type of substance—limestone decays faster than marble or granite—as well as its exposure to rain, sunlight, and wind. For marble and limestone, which tend to be the most common materials used in ancient monuments, acid rain dissolves the calcium carbonate that cements the stones' grains together, which results in eroded facades and structures.

Worldwide, structures that have existed for millennia can be heavily damaged by acid rain in less than a couple of decades.

Among the victims of acid rain are Stonehenge, the Tower of London, and Egypt's Great Pyramid of Giza. Acid rain may even prove to blur the faces on Mount Rushmore.

Other facets of global warming are also responsible for accelerated decay. One example of this is called salt weathering. Moisture on a stone structure causes salt to build-up in its pores. When the stone dries, the salt crystallizes and expands, which can fracture the stone. Scientists believe that this process will be accelerated as climate change produces wetter winters and drier summers.

Dr. Uri Dayan, who heads the department of meteorology and climatology, as well as the geography department, at Hebrew University in Jerusalem, says climate change not only is occurring in the region, but in all likelihood will be more extreme in years to come. This will result, he predicts, in more acute events, such as flash floods or sandstorms. The effects of severe weather events combined with significant shifts in temperature and more concentrated precipitation (rain and snow) will continue to do further damage to archaeological sites, ancient monuments, and historic buildings—so much so that in 2007 in its annual listings of endangered sites, the World Monument Fund for the first time listed climate change as a leading cause of destruction. The WMF says:

> The 2008 Watch List clearly shows that human activity has become the greatest threat of all to the world's cultural heritage, causing irreparable harm to many of the important places in the world that provide unique access to shared human history. Pollution eats away at ancient stones. The rapid rise in global tourism is bringing more and more people to fragile and often unprotected places. Cities and suburbs are spreading unchecked, at the expense of historic

landscapes and buildings . . . perhaps most daunting of all, the destructive effects of global climate change are already clearly apparent.

The 2008 list includes several sites that are threatened right now by flooding, encroaching desert, and changing weather patterns. "Sadly," it predicts, "future lists will bring many more."

Widespread flooding across Europe in 2002 destroyed an estimated 500,000 books and archival documents in museums and libraries. Floods attributed to climate change in northeastern Thailand have devastated centuries-old temples and Buddhist ruins in former capitals Sukhothai and Ayutthaya, the country's most famous archaeological sites. South Africa's West Coast National Park boasts artifacts from ancient humans, including the oldest human footprints made some 117,000 years ago. Archaeologists believe these and other treasures that remain undiscovered may be forever lost if the area succumbs to rising sea levels.

An estimated twelve thousand Scottish sites, including those that represent evidence of medieval salt workings in Brora, Sutherland; the Iron Age at Sandwich Bay, Unst; and the Vikings, at Baileshire, North Uist, are also threatened by erosion and sea level rise.

Acid rain has damaged more than 80 percent of China's thirty-three UNESCO-designated World Heritage Sites, including the Leshan Buddha. Rapid glacial melt in Huascaran National Park, Peru, is jeopardizing a cultural site containing pre-Inca treasures and temples that date back to 900 BC. Climate change is expected to bring even more flooding and likely further losses of historic and religious materials around the globe.

While the news of climate change might be focused on polar bears and arctic glaciers, we are also losing something vitally important to who we are. The past holds the keys to our shared human history as the WMF rightly notes.

I often hear people tell me that they don't think that their individual efforts can make a difference. Global warming is an overwhelming concept. But the truth is that every action makes a difference. We just need to start. We all have a stake in thwarting global warming and preserving our natural resources.

For example, turn up the thermostat in your home by one degree in summer and down by one degree in winter. If we all did it, we'd save an average of 8 billion kilowatt-hours of electricity and 160 billion cubic feet of natural gas. That equates to about 15 million tons of carbon dioxide in the United States alone, effectively eliminating an entire year's worth of emissions from three large coal plants. That plus the 25,000 tons of sulfur dioxide saved when we use less heating and cooling energy can help mitigate acid rain.

Around the world, wake-up calls are happening. Rome, London, and Stockholm all cleaned up their air quality when they began efforts to quell inner-city traffic congestion—and all the pollution that goes with it. Air quality can be improved by a quick 20 percent in cities just by limiting rush hour traffic. New York is also considering myriad ways to limit its inner-city traffic to combat pollution.

In Israel, environmental consideration of local sites is even bringing age-old enemies together. Far out in the desert, close to the border with Jordan, the Arava Institute works to bring Arabs and Jews together through environmental initiatives. "There is no place in the Middle East where Jews and Arabs work and study together for such a long period of time," David Lehrer, its director, tells me.

The institute hosts about forty students a semester and focuses on real issues, such as air quality, water and river pollution, and desertification of land. They try to figure out more efficient and sustainable uses for the limited resources they all share. "When we are working on real issues, common challenges, we don't have to talk about peace, we are doing something about it. We see the environment as a common cause," Lehrer says.

Within Jerusalem itself, Naomi Tsur, executive director of the Society for the Protection of Nature in Israel, is trying to preserve the historical sites, open spaces, and gardens so they will be there for the future. She is concerned about climate change because it damages the structures that give much of the world its emblems of hope. "Jerusalem," she says, "belongs to everybody."

When we examine issues through an environmental lens, consciousness is raised and the opportunity for change occurs. Ever walk past the same spot every day and never notice something particular about it until someone points it out? Take a stop sign. Every stop sign is exactly seven feet high. Its letters are all ten inches tall. And it is red because that is the color most historically associated with danger. It is a universal signal.

The environment has its own set of signals. Teleconnection patterns are what climatologists look at to gauge the world's weather traffic. These patterns show how weather in one area affects weather in another area. Technically, the patterns show large-scale changes in the atmosphere and jet streams. They show how temperature, rainfall, storms, and wind intensify over vast areas.

These patterns are often the culprits responsible for abnormal weather occurring simultaneously over seemingly far distances. For example, a winter that is very cold and snowy over much of eastern North America, will affect northern Europe and Scandinavia with increasingly cold weather, and make southern Europe and northern Africa very wet and stormy. These conditions are all partly related to the same teleconnection pattern.

There are ten common teleconnection patterns the Climate Prediction Center, an agency of the National Weather Service, routinely monitors. These patterns allow the Center to understand what is happening weather-wise in one place because of something occurring somewhere else far, far away. In other words, it shows them the connections of the atmosphere as a traffic map: buildup here creates a slowdown there, and maybe even conditions for an accident over there.

El Niño is the most well known example of these teleconnection patterns. El Niño is Spanish for "little boy" and refers to the Christ child because the phenomenon usually occurs around Christmas. El Niños are sharp surface temperature rises in the Pacific Ocean that lead to increased storm intensity. La Niña ("little girl") is the opposite of the El Niño effect, where surface ocean temperature sharply decreases. Whereas an El Niño would cause more rain in the Midwest, for example, La Niña would cause drought.

It shouldn't take a scientist to figure out that what we do in one place influences something else somewhere on the planet. If rain here affects snow "over there" it's easy to understand how acid rain over here could create acid snow over there too. That means pollution from a coal-fired power plant in Ohio can easily get caught up in the raindrops that fall on Massachusetts and further the erosion of Boston's historic monuments.

Despite the abundance of scientific evidence confirming the very real and dangerous consequences of pollution and carbon emissions on the global climate, there are still a handful of critics—scientists among them—who continue to assert that the realities of global warming have been grossly exaggerated by self-interested, left-leaning pundits and politicians. Let's set the record straight and get this nonsense off the table. The facts are irrefutable.

If you live in Kenosha, Wisconsin, or in Salt Lake City, Utah, you might be thinking that global warming doesn't make sense given that winter last year seemed to last longer than ever. So before we move on, let's clarify the difference between weather and climate. Weather is the current state of the atmosphere with respect to heat or cold, wetness or dryness, calm or storm, clearness or cloudiness. Climate is the average course or condition of the weather over a period of years. Weather is specific, whereas climate is more general. This means that in some places global warming, which is somewhat of an ill-defined term (a more accurate moniker for the issue is "climate change"), may indeed mean hotter temperatures. But in other places it may mean cold. The Earth tries mightily to balance extremes. On average, of course, the world's temperatures are rising due to climate change. In other words, snow on a mountaintop lasting to May doesn't mean that global warming isn't real.

NOW BACK TO finding that rock—the physical connector between all the faiths.

Next to the exit of the Western Wall tunnel, below the elementary school and meeting the surface of the Via Dolorosa, there is a brick of limestone. It doesn't stand out from the rest, but if there is a singular physical link to all three religions this is it.

I kneel, pretending to tie my shoe so I don't arouse suspicion from the Israeli soldiers keeping guard of the exit. I touch the rough edges of the stone and gray dust marks my fingertips.

The stone may not be part of the original wall. There may be a stone behind it that is the true epicenter of the world's religions. But does that really matter? What matters is that the stone is an extension of ourselves and what we believe it to be—a connection to something more ineffable, time defying. It's a point on Earth where we can, if we choose to, find connections to everything else in it. Limestone literally holds the material composition of our past. It's formed by the accumulation of organic remains over thousands of years. It's made of carbon, which, other than oxygen, is also the biggest element in the human body.

Caring for our future means caring for our past and understanding the basic components of our existence—air, water, the food we eat. But importantly, understanding needs to enter the equation; it is the missing link in the ecology of the planet.

The environment doesn't have to become an issue of biblical proportion. We can take steps to preserve it once we've determined that we actually do care about it.

We are all connected to history and that should embolden us to preserve the world for future generations. However, there are some places where it may be too late. In the next chapter we visit what many say is a hopeless cause: Mumbai, India.

Our Future

Mumbai, India

The sunset in Mumbai draws people to the shore much as it does in Key West in winter or California in summer. Although the crowd is a little different: Muslims in skullcaps and burkas, Hindus wearing red-dot bindis between their eyes, and Indians in business attire all take up their places on the wall that overlooks the Arabian Sea. Mumbai was called Bombay until 1996.

A monkey on a leash scrambles by. The ubiquitous scrappy dog found in every Third World country and a goat check each other out. A giant ox pulling a cart storms along. Then a Lamborghini zips through traffic, beeping its horn.

Mumbai is one of those places that you have to see to believe. It's an orgy of color, mayhem, flash modernity, and squalor. It's what happens when hordes of people are forced onto a narrow patch of land and urbanization takes hold.

It's wondrous. It's tragic. It's our future if we aren't cautious.

Mumbai has one of the largest urban populations in the world; Greater Mumbai has almost 20 million people. Although India remains 70 to 80 percent rural, people come to the city because there are more resources to sustain them. Mumbai itself may be poor, but in the city at least there are possibilities for water, food, and earning money.

More people are being drawn to urban areas all over the world because of these possibilities. And these urban areas are increasingly coastal. Coasts provide easier access to trade and transportation. Trade and transportation beget money and goods. Coasts are where most markets are created.

Just twenty-five years ago less than 2 percent of the global population resided in "megacities" of 10 million or more inhabitants. Today, the proportion exceeds 4 percent. And by 2015 it will top 5 percent, when megacities will likely house 400 million people.

Even in the United States, coastal urban areas are on the rise. More than half of the US population lives within fifty miles of the shore. Coastal cities have swelled by almost 30 percent over the past twenty years. And all indicators point to even more people moving closer to the water.

Mumbai is a case study in the problem of larger centers of populations moving closer to water. Greater Mumbai is low-lying and therefore flood prone. There are virtually no exit routes from the city, so the people are captive. Mumbai is infamous for its pollution, inadequate landfills, hazardous industrial wastes, and rampant diseases. Sewage in the city is not treated before it is discharged into the Arabian Sea. All sewers overflow into coastal waters adjoining Mumbai, which make the waters unfit for marine life or human use. Hundreds of septic tanks overflow into the ground, causing flies and mosquitoes to breed. Millions live in slums, where respiratory disorders are common, and gastroenteritis, tuberculosis, malaria, and other parasitic diseases like filariasis are widespread. Every day, 880 million gallons of drinking water must be brought to Mumbai from a distance of over one hundred miles. Two million people live with no toilet facilities.

In short, Mumbai, already operating in a precarious state of existence, will see things deteriorate further. By 2050, the city will experience a temperature rise of between 2.25 degrees and 3.25 degrees along with an average annual decrease in precipitation of 2 percent, according to climate model estimates. The Canadian Climate Center is predicting Mumbai to have a nearly 20-inch sea level rise by 2050, which would prove disastrous—the city would be a total flood zone. Meanwhile, the increased temperature and the lack of rain would yield additional serious consequences to a place that relies heavily on rainwater and whose heat attracts strong storms. Just a few years ago more than a thousand people died when a monsoon struck.

Mumbai comprises seven islands that are between thirty-three feet and forty-nine feet above sea level, making it virtually defenseless against any sea level rise at all.

Now consider this: at its lowest average point, Mumbai is at the exact same elevation as New York City. The problems that can occur in Mumbai can just as easily occur in other major cities around the world; sea level rise doesn't discriminate.

Currently Mumbai floods rapidly when heavy rains combine with high tides. Storm drains are often blocked by trash and debris. Monsoon flooding frequently shuts down Mumbai's commuter rail lines, which are among the most heavily used in the world. With about 100,000 people on every square mile, its squatter communities are the most densely settled districts on the planet. The World Bank estimates that 170 persons depend on each public latrine in these areas, and that one-third of the 35,000 latrines in Mumbai are out of service. Without facilities, people use the ground. And when it floods, that creates disease.

Landslides are another threat.

It's a lot to take in. The problems seem endless. It's a lot easier to just turn the page and ignore the plight. Except you can't turn the page. A good portion of us are living with the threat of similar consequences. For example, one-tenth of the world's population lives in low-lying coastal areas like Mumbai. Of the 180 countries in the world with low-lying coastal zones, 130 of them, including the United States, have built their largest cities on these highly vulnerable lands.

The ten countries with the largest numbers of people living in these low-elevation zones include (in descending order) China, India, Bangladesh, Vietnam, Indonesia, Japan, Egypt, the United States, Thailand, and the Philippines.

Debi Goenka, one of India's most well known environmentalists and head of the Conservation Action Trust, says the world is starting to get it now, albeit at Mumbai's expense and example.

"Climate change has woken up a lot of people. Finally, something seems to have clicked in people's minds and they are realizing that it's all linked. Before, environmental issues were compartmentalized. Now people are making the connections," he says, linking urbanization to water shortages; temperature rises to stronger storms; increased flooding to disease; lack of recycling to more waste; more waste to the need for more natural resources; more natural resource demand to more intense material extractions in rural areas; resource extraction to the displacement of rural populations into urban areas; urbanization to . . .

When I meet with Debi at his home in northern Mumbai, he waxes eloquently about the issue of global warming, but I notice that he is speaking around the issues that have to do with Mumbai, but not specifically about Mumbai. I point this out and ask him why. "I'm afraid," Debi explains, "that the time has passed for Mumbai."

Debi is fifty-something. The gray tips of his hair meet the frames of his glasses. He speaks low and fast and it is difficult to hear him sometimes, which is frustrating, as he is precise and austere with his words. "My goal," he says, "is to get people to believe that doing something is enough, that every little bit helps."

It takes me about ninety minutes to get to Debi's apartment twenty-five miles away from where I am staying—at a hotel in front of that seawall on the water. The traffic is brutal everywhere in Mumbai. There are more than seventy thousand black-and-yellow taxis and autorickshaws that sway bumper-to-bumper in lanes of six that are meant for two. Somehow it all works out in a hot, cacophonous flow of steel and machine and alertness. The slightest veer from one driver sets off a ripple of reaction. Drivers beep their horns every few seconds. It has to be this way; otherwise all would stop still in a muddle of apprehension. It brings to mind the VISA advertisement where one person pays cash and the whole system stops.

When you gaze at the wonder of sights in Mumbai you see temples, mosques, architecture, and designs that remind you at every single turn that you are not at home. The colors and landscape defy familiarity. Then you look beyond the fashion, the customs, and the culture and look at the people. They are walking to work, going to school, shopping. They are us.

Shaking off the hectic ride back to my hotel from Debi's, I stroll across the street to the seawall. It's fifty steps from hotel door to the ledge. In front of it is a giant concrete barrier. The barrier is made of huge stone jax—six-pronged concrete pieces piled up on top of one another in the water. They are ominous reminders of just how close Mumbai is to losing itself to a storm.

"The sea has a funny way of reclaiming what was once its own," Debi so poetically explains. Mumbai is built on reclaimed land.

Mumbai itself is like an epic poem. Its roots lie as a fishing village dating back to the second century BC. Then Muslim sultans took over the land, annexing it to the mainland of India. In the

1500s it was taken over by the Portuguese—mighty sailors at the time—who named it Bom Bahai (beautiful bay). They gave the city to the English as part of the royal dowry of Catherine of Braganza when she married Charles II in the mid-1600s. The islands were then leased to the East India Trading Company, which built a fort upon the land and began a reclamation project that lasted sixty years to make the islands one landmass. The newly named Bombay flourished as a port and became the largest distributing entrepôt in India during the 1800s after the Suez Canal was built. It was a central city in India's independence movement from Great Britain and is where Mahatma Gandhi staged many campaigns. In 1996 the city's name was changed to Mumbai after the Hindu goddess Mumba, whom the original fishermen of the land worshipped.

The simple things you see and hear have meanings upon meanings. When I walk to the Gateway of India, a huge archway in front of the Taj Mahal Hotel on Mumbai Harbor, I see the tourists mill and snap photos. But beyond, way in the distance, are convoys of shipping tankers. They are low riding, meaning they are full. And they are making their way into port filled with goods . . . and waste, often hazardous waste.

This is where the traces of American lives, thousands of miles away, can be found steaming to shore.

About 80 percent of the electronic waste in the United States is exported, mostly to Third World countries like India. It comes by ship, on tankers filled with used computers, cell phones, televisions, batteries, all kinds of things that contain mercury, lead, and heavy metals that are dangerous to people's health and the planet.

In India the hazardous waste isn't treated properly, nor is it disposed of with the right type of environmental consideration. But it's cheaper to get rid of it here than in the United States.

Toxics Link, an India-based nongovernmental organization that tracks electronic waste (or e-waste) in India, has calculated that it costs about twenty dollars to recycle a personal computer in the United States, whereas unscrupulous Indian importers pay up to fifteen dollars each for them. That means a net gain of thirty-five dollars for a US recycler. By extracting the usable parts and then dumping the rest on backyard scrap-trading outfits, an importer can generate about ten dollars in revenue. Meanwhile, we pawn off our hazardous material onto people who can least afford to have it dropped on their doorsteps.

The two largest nations exporting their e-wastes are the United States and the United Kingdom. According to a recent British Environmental Protection Agency report, Britain shipped out twenty-five thousand tons of e-waste to South Asia in 2005. The United States exports ten times that amount annually. For every PC we buy, we discard one. While about two billion dollars worth of electronic equipment is recycled in the United States, it represents just 11 percent of the e-waste generated. Put into simple terms, the tonnage of e-waste we toss out with the garbage each year is equal to about half a billion laptop computers.

Worldwide, about fifty million tons of electronic waste alone is generated per year, although what is exported isn't labeled hazardous waste. Waste is considered "hazardous" if it contains corrosive, toxic, ignitable, or reactive ingredients. Hazardous waste includes computers, batteries, cell phones, and other products that contain toxic chemicals. It usually goes by ship, and isn't labeled as waste because it's largely illegal to ship hazardous waste. Instead, Dr. Kishore Wankhade, the regional coordinator for Toxics Link in Mumbai, says corporations and waste management middlemen will falsely label containers, sometimes even as "charitable goods." India bans the importing of "waste," but the business is so profitable that the Indian government is considering changing the language of the law to read "hazardous material." This would recategorize waste and make it allowable as imported "material." "Through a not-so-subtle mangling of international definitions for *waste, disposal,* and *safe recycling,* the Indian government has designed a veritable global waste funnel that will ensure that the world's waste will surge to our shores," said Ravi Agarwal, director of Toxics Link.

"Already, four thousand tons of e-waste is piling up with no systems in place, no collection, no recyclers, or anything," Kishore says. He points to an old desktop computer of his that sits in a corner. "I'm waiting," he says. "I want it to be on the first recycling program when it comes." He admits the computer will likely sit there for years.

It isn't just India that suffers this plight. In Lagos, Nigeria, there is a landfill site they call Computer Village because it is stacked high with machines from all over the world.

The Basel Action Network, a Seattle-based nonprofit that works internationally to ban toxic trade, has examined the waste, which isn't treated or filtered, and found parts from as far away as Los Angeles. Indeed, on its Web site and in a film BAN produced about e-waste, you can clearly see the marked labels of the Los Angeles County School district. BAN says:

> There is an ugly underbelly of economic globalization that few wish to talk about. Under the guise of simply utilizing the "competitive advantage" of cheap labor markets in poorer areas of the world, a disproportionate burden of toxic waste, dangerous products, and polluting technologies are currently being exported from rich industrialized countries to poorer developing countries. In effect, rather than being helped to leap-frog over dirty development cycles directly toward clean production methods, developing countries are instead being asked to perpetuate some of the world's most

toxic industries and products and are even asked to become the global dumping ground for much of the world's toxic wastes.

California and most other states allow the export of hazardous waste. If you've tossed out batteries, cameras, lightbulbs, or anything electric that contains toxic chemicals, likely it's now somewhere in the developing world.

The problems begin when the parts of those electronics begin to decompose.

The Silicon Valley Toxics Coalition reports that just one computer can contain hundreds of chemicals, including lead, mercury, cadmium, brominated flame retardants, and polyvinyl chloride (PVC). Many of these chemicals are known to cause cancer, respiratory illness, and reproductive problems. They are especially dangerous because of their ability to travel long distances through air and water and to accumulate in our bodies and the environment. And that's why we end up shipping e-waste far, far away to countries halfway around the world.

To be sure, it isn't just foreign waste that ends up in the landfills in India. Economic growth is causing domestic waste to increase as well.

India's computer inventory is growing at quick clip, according to Kishore. He says right now there are about eight to ten computers per one thousand people in the country. That is expected to increase to fifty computers per one thousand over the next few years, and "with the kind of population India has [more than 1.1 billion people], it will be a big problem," Kishore notes. It also means that more computers will get tossed faster.

By example, Kishore says the average life span for a computer in India was seven years—until two years ago. Now the average life span is five years, and shrinking quickly to three, he says, as corporations trade in their hardware for new models.

In the United States, the average life span for a computer is less than two years.

If e-waste exports are banned, as Greenpeace, BAN, and the Silicon Valley Toxics Coalition, among many other groups are lobbying for, we'll see an acre pileup of computers and electronics more than 1,500 feet high, eclipsing the Empire State Building and causing a health hazard unlike any the United States has ever seen.

E-waste is such a huge problem in India that I decided to see it for myself. I traveled to one of the largest landfills in India, on the outskirts of Mumbai, and waited in line to enter in a small, white sedan, with massive Clean Up Mumbai dump trucks in front of and behind me.

Hundreds of people live on the grounds of the landfill in a small shanty village. As I waited, villagers came and went from a bakery, hauled wheelbarrows full of trash here and there, and bustled about. A goat looked up at me in my car, wasn't interested, and went back to eating the garbage that had spilled out of a plastic bag on the ground.

Past the landfill's entrance gate I was stopped by a security guard and decisively removed from the premises, which was all right by me because the smell was so bad I probably would have passed out soon enough anyway. It smelled like spoiled milk and excrement, and as if you were locked in an airless closet with that heated combination festering.

Sorry about the graphic description, but I detail this because I want to put you there and remind you that a village of hundreds lives in this place. Visually, as far as I could see were piles and piles of refuse. I didn't witness any separation efforts, as is the case in US landfills: the paper here, the plastic there, food yonder, metals pushed neatly to the side. I've been to my share of landfills in the United States. Here it was one big pile.

"There used to be a recycling program a few years back," Debi says, "but the municipality would just pile everything together anyway so people gave up."

Kishore says hazardous waste ends up at the landfill I was at, hospital waste too. "They make it look as though they are treating it, but they aren't," he explains.

Because it is expensive to treat and dispose of hospital waste properly, many will toss some of it out with regular waste, or say they are burning more than they are—and then toss the rest. Burning hospital waste is a common means of disposal even in the United States. Air pollution becomes a whole other issue.

There are so many issues to focus on, Kishore says, that sometimes it is difficult to focus. He says his most immediate and biggest goal is creating awareness among people about the ramifications of their actions.

"People don't think the environment has anything to do with them. We have to educate them so it comes from the inside. It should not just be about cost cutting. Health, water supply, and education are connected," he says.

We need only look at Mumbai to see what the world may have in store. All of the environmental issues it is experiencing can be found in almost every country in the world, just at lower or less progressed levels.

THREE CHILDREN NO more than ten years old come up to me at the Deonar landfill and just stare. One has a flat stick, and another a rubber ball. They were playing cricket—a far cry from the whites and neat lawns you see associated with the sport on television.

No one, especially children, should be relegated to such a life whereby they have to play on piles of refuse, breathe toxic pollution

created by hazardous waste, and believe that this is life, this is just
how things are in the world. I want to be able to explain to them
how life can be—life as we experience it in the United States without a thought: clean water, air, and spaces on which to play. Then I
think about our carelessness in the United States. I want to be able
to explain to Americans too that this is how life can be: life lived
on the ousted remains of our days.

At a recycling center just a few kilometers from the landfill, an
entirely other existence takes place. For miles, trucks line up in
front of shanties specializing in certain materials: steel, plastics,
cloth, stones, cardboard. They buy the refuse, clean it, and sell
it off.

When you drive by the site it's incredible to see just how vast
the center is. Yet you have to wonder, what's the point? All of
Mumbai looks like it's in need of recycling.

"If we didn't recycle," Debi explains, "we'd drown in our own
waste." Also in a simple example of reducing, Debi says that no
one in Mumbai would ever, as they do in the States, take a few
sips from a can of Coke and toss it away. "We just couldn't comprehend wasting it." They use everything until the last drop, he
says, until waste is forced.

Since I was born forty-one years ago, the amount of waste
every American individually produces has nearly doubled, from
over 2.5 pounds of trash per day to 4.6 pounds today. We recycle
about one-third of that. What we toss the most is paper and
packaging. Think about all the scraps you end up with when
you buy anything these days. I just opened a box of cookies, for
example. I had to peel off the sticky labels, rip off the cardboard
tab on the box, and then take the cookie out of the plastic wrap-
ping inside the box before I could eat it. We use 164 million
plastic bottles, 273 million aluminum cans, and 350 million
paper cups—per day in the United States. Only about 7 percent

of all plastic in the United States is recycled. While we recycle 88 percent of newspapers every year, we still leave more than 1.5 million tons discarded. Our total annual waste amounts to 169 million tons. Without recycling it would amount to 251 million tons a year. One million tons is the weight of a freight train with ten thousand boxcars.

Recycling isn't much of a choice anymore.

Then there is the e-waste that is shipped out of the country. We produce about three million tons of e-waste per year plus enough household hazardous waste—batteries, paints, light-bulbs, etc.—to fill the New Orleans Superdome 1,500 times over. Think about what would happen if we didn't export our waste and had to live with our own.

The picture of Mumbai starts to look more and more familiar.

Still, exaggeration and alarmism are too frequent in environmental writings. Like everything, the conditions that plague Mumbai are complicated and consist of many factors, not just one or a few. We are, after all, talking about a society that has existed for five thousand years. But there are parallels and there are consequences, which are very real. We can take preventative steps—even little ones count.

Debi points to the little things people in India can do that make a difference. With more than a billion people, results can happen fast. He talks much like we do in the States: turn out the lights, recycle, take public transportation. But what gives me the most hope for change in Mumbai is something tragically ironic.

What is going to save Mumbai and maybe all of India from collapsing from environmental destruction are the slums.

In Dharavi, the largest slum in all of Asia, I saw hope crawl out of the most desperate of places. I witnessed an amazing economy built on the scraps of society. In shanties and shacks, people accept, clean, and then reprocess the junk discarded on roadsides and thrown away en masse from corporate dealers looking to offload their worn and used goods.

It's about an hour's train ride to the Dharavi slum from the Churchgate Station near my hotel. Dharavi is famous in Mumbai because it is so big, not in terms of land—it's only a square mile—but in terms of people: more than a million people live there, crowded between railroad tracks. There are permanent homes, some going as far back as 1969, and shops dealing in different scrap materials. More than two thousand products trace their roots back to Dharavi. It's well known for its leather products, pottery, and baskets.

The economy that Dharavi has built is estimated to produce more than $650 million per year in revenue. Make no mistake: it's a slum of epic proportions. The average resident earns just a few hundred dollars per year, and 40 percent of the population earns far below that, according to local reports. Slum owners are rumored to take most of the revenue that is derived from Dharavi.

"The slum owners come every day in their Mercedes to check on things. The workers [meanwhile] cannot even afford a train ticket," Sunil, my twenty-two-year-old guide into the bowels of Dharavi tells me. My train ticket cost me thirty cents each way.

You can't just show up at Dharavi, walk around, and take pictures. You need permission to go there. Diligent research gets me the contact I need to get in: I find a tour service in Lonely Planet's *India Travel Guide*.

Before you think less of me, let me explain. Reality Tours and Travel is a nongovernmental organization that donates 80 percent of the proceeds to Dharavi. It sponsors training, English, and education classes. And you don't walk around pointing as in a zoo and gushing: "Look! A slum dweller!" Reality Tours points out the positive aspects the slum brings to Mumbai and India's economy in general. It showcases the businesses and the people who work very hard just to survive; they are living off, from, and due to recycling.

Sunil meets me and four others at the train station to escort us to Dharavi.

Lisa and Ellen, two young British women who could easily be mistaken for the Pigeon sisters from *The Odd Couple,* are standing at the ticket window looking distinctly out of place. Bird number one is wearing flip-flops and thin linen pants calf high and butt low. Bird number two is in an ultra mini skirt. Underneath, thank God, are black tights. They both sport matching brown T-shirts. I'm guessing they are twenty-something. James and Eloise, the two others, are also Brits. They're in their thirties and have been traveling in India awhile. James has a red tilak mark of ink between his eyes, which looks out of place because he is wearing a baseball cap, polo shirt, and jeans. Eloise, meanwhile, looks exactly like you'd expect an Eloise to look like, round rim glasses and all.

Just as Sunil greets us, an old man comes by and inquires about what we are doing.

"A tour?" he says, as we explain.

"Dharavi," he repeats.

"A lot of people don't like the fact that we are pointing out Mumbai's slums," Sunil explains. "Until we tell them why we are doing it."

Reality Tours and Travel, he says, was begun two years ago and has caught on. Sadly, the business has a future.

On the train ride, Sunil fills us in on the no-photo and other "no" rules: no shaking hands, no giving money, and no stopping; we must keep moving at all times. "Some people there," he says again but more emphatically, "really don't like us doing this."

Corrugated steel patched together. Dirt roads. Tiny shanties. I'm warned to duck my head often because the structures are built so low and their overhanging roofs jut out.

Our first steps in to Dharavi bring a team of kids to greet us. They come right up to us with their hands out. Sunil looks and shakes his head. The team retreats behind us, but follow us the whole time we are there.

Imagine taking four pieces of wood, six feet by six feet and somehow attaching them together with nails, boards, and scrap metals. Then put a piece of corrugated steel on top and you have the basic makings of a structure in Dharavi. Then attach other structures to three of your walls and there are your neighbors. Streets upon streets of mixed and matched shacks like this make up the core of Dharavi.

The outer buildings are where the shops are. There, rag pickers come with piles of different things wrapped in keg-sized burlap sacks or webbed plastic. They sell the junk contents to the slum owners. Light sockets. Paint buckets. Oil drums. Plastic bags. Cardboard. Computers. Plastic cups. Plastic plates. Plastic bottles. Glass of every shape and size. Detergent bottles. Pots. Pans. Soaps. Liquids. Clothes. Dishes. Smoke alarms. Almost anything you can think of ends up here.

Next, workers separate the good stuff from the bad stuff and clean it. The whole bunch then gets put into the reprocessing cycle out of which comes new material. The material is then sold to corporations out of which new products are made . . . and sold to, well, us.

Belts, wristwatch straps, and wallets are big exports from Dharavi.

The first stop on this miraculous tour is the computer shop. Stacks of desktops pile high against a wall outside. A worker takes the housing off a monitor and brings it inside to other workers at a shredder. The dust flies as the plastic case is chopped into pieces. These pieces are then put into a press and rolled together. After that, they are burned, smelted, and put through a sort of colander, something like one of those Play-Doh presses that produces strings of material. The plastic strings are then dried and sliced into pellets. Those pellets are sold to companies that use them to make new computer cases, or other products such as toys.

"They sell without regard to the toxins," Kishore later informs me when I ask him about the plastic products made in Dharavi. "Sometimes the companies will mix that plastic with virgin plastic to lower their production costs. No one knows. They are able to skirt regulations and toxic standards this way. Or else they just go ahead and make new toys of it, even with all that lead, and sell them on the informal market."

The toymaker Mattel, as was widely reported, got into a heap of trouble for selling toys in the United States that had traces of toxins in them. This was a small amount and an aberration. But on the informal market, a market of cheap goods that caters to poor people, those who cannot afford brand-name toys or goods, the practice of using tainted materials is rampant, Kishore says.

I'm shocked to see the whole computer disassembling process occur before my eyes. It's one of the things I wanted to learn more about on my trip to India. I never thought I'd get to see the process up close.

The toxic dust, the lead, mercury, and other materials are laid to the side to decompose. When plastic decomposes, remember, it emits an array of toxins and dioxins into the air that cause all sorts of diseases and health hazards.

I know I said this section was going to be more positive. In a relative sense, it is. There is work. There is money to be made. People aren't dying of starvation as they likely would in the rural areas. The recycling process keeps the junk from overtaking Mumbai. The consequences, though, can be dire.

Some regulation and oversight would go a long way toward making Dharavi a real force in the economy. And such regulation may not be that far away.

The government is looking to take over Dharavi, raze it, and regulate the businesses that have mushroomed there. They'd provide housing to people who have lived there since before 1995, Sunil says. And they would clean up things. Of course, more taxes would be levied. Still, that might put the slum on the straight and narrow.

I saw paint cans and oil drums dumped, washed, and banged back into shape, the contents of which drenched the ground all around and near the shops and homes. Sniff an open paint can for just a second and you get an unforgettable toxic whiff. Multiply that by a hundred and then live with it day and night. That is what it is like here with the fluids spilled all over the ground. Slum owners don't care; they don't live there. That's why government intervention may be good.

Also, it might stop knockoff goods from being produced there. Want a laptop case? Select one you like there, and a logo shop next door will slap a Sony label on it.

"Reebok, Adidas, anything you want," Sunil says.

If you own an item that is forged, likely it began its life in a place like this.

The alley is narrow and dark, and I have to bend real low to keep from banging my head. The Pigeon sisters gaze about in wide arching glances, their heads rotating like machine guns on turrets as they blabber in their high pitched East London accents. "Oh, looky there, and there! That's straight out of Leicester Square!"

Eloise blinks like an owl at the entrances of homes shrouded by cloth, some with ladders outside that lead to even more rickety second floors. Eyes peer out. James bangs his head badly on a pipe that comes out of nowhere. Sunil keeps us marching; he doesn't like this area. And, he says, in this area the people don't like him. He asks me to lead the group along, single file, through the alleys as wide as my shoulders.

I have no idea where I'm going, of course, and can't hear Sunil's directions. He is the last man in line.

I take a wrong turn and lead us into an open doorway where a man is changing his pants. Eloise blinks madly. The man doesn't flinch. He goes about his business.

We about turn and end up behind a row of shanties. This is where the raw sewage collects. I gag. This is also where the slum dwellers dump their own garbage and trash—and burn it. Burning trash emits ten thousand times more pollutants than if it were sent to a proper landfill. Those pollutants may include dioxins, formaldehyde, hydrochloric acid, and ash containing arsenic, mercury, and chromium, among other toxins. But that is of little concern to the people who live here. "How do you explain that concept to people who are looking to just survive on a daily basis?" Debi reminds me later in conversation. "They cannot afford a long-term view."

The group of kids that has been following us catches up to us here. They again want to shake hands. Sunil shakes his head. But I look into their eyes, and I extend my arm.

Electrical, plumbing, and art classes are beginning in Dharavi, sponsored by the Reality Tours' NGO. And the Dharavi Redevelopment Project launched by the government is getting underway. It could be a bright example of that which can be raised from the dust—literally—of existence.

Education is arising in a place where the need for survival forced opportunity into being.

There are more than two thousand slums in Mumbai taking back what was once ours and recycling it into something new.

Without the slums of Mumbai, Debi says there would be no Mumbai.

The Pigeon sisters, James, Eloise, and I part ways without much fanfare. How do we connect with each other about what we've just seen—human beings, people in the big scheme of life, just like us living like that? Going on the Dharavi tour wasn't like watching a soccer match together, relaxing after a day of skiing, or some other common experience that breeds individual remembrance and points of view. Dharavi was an experience of spirit, and we each had to rest with our own.

"The noblest conception on earth is that of men's absolute equality."

The solution to the dilemma of life in Mumbai's slums—as well as to our implicit role in contributing to it—lies in individual compassion.

Water. Waste. Energy. Everything in the world can be traced back to those three things.

In Mumbai water is scarce.

In Mumbai waste is rampant.

In Mumbai there's an energy shortage.

This is the cost of urbanization, of growth; we step on each other, selfishly, mostly unconsciously. We've yet to realize that the global village people have talked about for so long is here. Now the question is, how do we all live together and make the best of it?

And humans are just one of many species. What we are doing to all the other living things on the planet, we will see in the next chapter.

THREE

We Are Not Alone

Borneo, Southeast Asia

I trekked through the jungle today to see firsthand the logging camps on the Indonesian border. It wasn't easy getting anybody to take me.

"Logging camps? No, too dangerous."

"Logging camps. No, how about a cultural tour?"

"Logging camps. That's a one-way ticket."

I'm not here in Malaysian Borneo to visit wildlife preserves, take a sunset cruise, or to see the locals put on some touristy song and dance. I want in—way in—to the jungle to check out what's destroying it and the species here.

Borneo is home to the most species-rich places on Earth.

No one in Kuching, the most populated city in this part of Borneo, where I am staying, including the World Wildlife Fund (WWF), will escort me.

Loggers, people around town tell me, are largely gangsters. They hire Indonesian workers for pennies, scam people, allegedly pay off government ministers, and log in protected areas. No one wants to get near them because they also reportedly bring crime and prostitution, the usual trappings of the underworld. They are dangerous.

Since no one in Kuching is willing to take me, I book a ticket to another town, Sibu, nearer to the center of Borneo, where logging is more rampant. In Sibu I hope to rent a speedboat that will take me downriver to some logging camps. This area is regarded as the starting point for the journey to the heart of Borneo, otherwise known as the deep, deep jungle—one of the most exotic places on Earth.

As I'm packing the night before I'm set to leave on a boat that will traverse the South China Sea and bring me along rivers lined thick with mangroves, the phone in my hotel room rings.

"Thomas," the voice says, "I have good news for you. I have found someone to take you to a logging camp." The voice is of a shady local who said for the right price he might be able to help me find evidence of deforestation in the jungle.

It's several hours from Kuching by car, and then another several hours through the jungle on foot, I'm told. But first I have to hire a team of people: an interpreter, and two Iban tribesmen to guide me through the jungle. I'm quite sure my tipster is receiving kickbacks from all the people I hire.

Wading waist-high through rivers, stepping thigh-deep into mud along relentless jungle trails, up hills that have me wheezing, I trek for miles.

Across a fast-running river, an artificial hole has been chopped through the trees. I hear a chainsaw, the first man-made sound that drowns out the birds, bugs, and whatever that thing is that makes a high-pitched whooping sound.

I step onto the grounds of the logging camp.

I am one of the few outsiders who ever has.

— — —

BORNEO IS AN island in Southeast Asia that is divided between Malaysia, Indonesia, and Brunei. It has 15,000 species of plants, 3,000 species of trees, 221 species of mammals, and 420 species of birds. Just one small patch of forest here can hold as many species of tree as there are in all of North America. But here is where they log and fell trees, to the tune of billions of dollars a year. Malaysia and Indonesia are the biggest tropical timber exporters in the world, accounting for more wood shipments than Africa and Central America combined.

That isn't all.

When the trees are chopped, the land is then used to grow palm oil. Malaysia and Indonesia are the largest exporters of palm oil in the world. Palm oil is the most commonly used edible oil after soy. It's in your toothpaste, your breads, your soaps and your cereals. It's often just labeled "vegetable" oil, but its use is widespread. We don't much notice it even when it's staring us in the face. For example, ever wonder where the name Palmolive comes from? Palm oil.

But the people, animals, and plants on Borneo certainly notice how much palm oil we use.

"It's grown all over," Richard, my interpreter on my jungle trek says. "Look." He points to a patch of land just to my right off the trail. "Palm."

It wouldn't be such a big deal, all this palm oil production, except that it is ravaging the land and displacing species, in some cases causing them to become extinct. Here you can find orangutans, monkeys of different varieties, tigers, leopards, rhinoceros, elephants, wild boar, deer, dolphins, all sorts of fish, and thousands of plant species, with some still being discovered. It is a virtual Eden.

According to the World Bank, the lowland forests of Indonesian Borneo will be completely cleared by 2010, and the upland forests by 2020 due to palm oil plantations and timber cutting.

We are condemning the wildlife here; so much so that while I was there Greenpeace was staging a blockade of a ship offshore. With a banner reading "Palm Oil Kills Forests and Climate," the Greenpeace vessel positioned itself close enough to the tanker to prevent the tanker from leaving port. The tanker was filled with 33,000 tons of palm oil.

"The expansion of palm oil plantations into forest and peatland areas poses a serious threat to the global climate and Indonesia's remaining forests. Expansion plans in Riau province alone have the potential of triggering a 'climate time bomb.' Riau's peatland forests store a massive 14.6 billion tonnes of carbon—equivalent to one year's global greenhouse gas emissions," Greenpeace claims.

Activists regularly stage rallies and campaigns against the use of palm oil. NGOs lobby the government and try to establish protocols for sustainable farming pretty much to no avail. The Roundtable on Sustainable Palm Oil says,

> Driven by ever increasing global demand for edible oils, the past few decades have seen rapid expansion in the production of two major edible oils, soy oil in South America and palm oil in the tropics. From the 1990s to the present time, the area under palm oil cultivation had increased by about 43%, most of which were in Malaysia and Indonesia—the world's largest producers of palm oil. While better managed plantations and palm oil smallholdings serve as models of sustainable agriculture, in terms of economic performance as well as social and environmental responsibility, there is serious concern that a significant amount of palm oil is not produced sustainably.
>
> Development of new plantations has resulted in the conversion of large areas of forests with high conservation value

and has threatened the rich biodiversity in these ecosystems. Use of fire for preparation of land for oil palm planting has been reported to contribute to the problem of forest fires in the late 1990s. The expansion of oil palm plantations has also given rise to social conflicts between the local communities and project proponents in many instances.

WWF-Malaysia takes a different approach to sustainability in Borneo. It encourages palm oil production, but lobbies businesses to adopt best practices to foster responsible farming.

Junaidi Payne, WWF's technical adviser to species and forest-related issues in the Malaysian state of Sabah, just north of where I am in Sarawak, says boycotting palm oil altogether is a stupid idea. "It's just unreasonable," he says in a British accent laced with pauses of "mmms" in between. "Most NGOs think oil palm is the enemy but it isn't. The issue is government policy," he says.

Most of the laws governing forestry preserves in Borneo were set in the 1970s.

At the time, the government was focused on making big money from logging, and sold off about half the land to private individuals and kept the other half for themselves.

There has been a struggle over the past two decades to transition the Malaysian economy from agriculture to manufacturing in order to keep the economy viable. As the amount of jungle forest available to log dwindled, the government needed a new source of income. Along comes palm oil, because it's cheap and relatively easy and fast to grow. A new source of funds was found. The policy again, Payne explains, wasn't to protect and conserve but to farm, manufacture, and profit. The fact that palm oil has become the source of these profits is incidental. "It could be rubber or rice; it doesn't matter. It just happens to be palm," he says. That's why a total boycotting of palm oil won't change the conditions in Borneo, it will merely replace the crop. Then what? Boycott rubber and rice, or whichever substitute crops the private landowners and the government find to replace it? No, Payne argues, better to create more environmental awareness and education and foster sustainable farming than to swap perversities.

"WWF is supporting the government through a joint program of wildlife conservation to make the forest productive again," he says. It is restoring the forests by planting fast-growing natural, native species that will allow logging and farming but won't decimate the land for generations. Planting faster growing species of flora obviously allows a rate of regrowth that is more expedient, so plants and trees can be harvested but will grow back more quickly.

Losing forests—which is what deforestation means anyway— is the biggest man-made contributor to climate change after burning fossil fuels. Deforestation accounts for about 20 percent of global carbon dioxide emissions per year, mostly resulting from the burning of forest residue after an area has been logged. This is only half the story of how deforestation contributes to climate change. Trees naturally remove and store carbon dioxide from the air. So less forested areas mean more greenhouse gases stay in the atmosphere, which further contributes to global warming.

Based on data provided by the World Bank and the UN Food and Agricultural Organization, deforestation in Borneo is responsible for an estimated 715 million tons of carbon dioxide per year, equal to the annual emissions from over 120 million automobiles.

The Malaysian states of Sabah and Sarawak are the areas most pillaged by logging and palm oil production. Brunei, which rests between these two states, is relatively immune because it is small and runs largely off its oil and natural gas reserves. Indonesia, on the Malaysian border, is also a big exporter of palm and timber and is famously reckless in its environmental practices.

The WWF and other organizations cite Indonesia's track record of environmental responsibility as atrocious. As much as 90 percent of the timber harvested from Indonesia is reported to be from illegal logging—operations in wildlife preserves or in areas that are not regulated.

It isn't very easy to get the world to wake up to the trials and tribulations of Borneo either. The orangutans can't get together and stage a rally for their survival. The elephants can't campaign for more space. Even the dolphins and aquatic species are being damaged by the production of palm and timber refining because of the refuse spilling out to sea.

More than 120 million people may want to start thinking about the environmental troubles of Borneo, because that is how many people rely on its fisheries for food and its plants for vitamins. Those fisheries are fast being emptied. Because of the exploitation of Borneo's natural resources, aquatic systems are polluted and the food chain is breaking down. The area's coral reefs are becoming endangered. As the WWF explains, "If our oceans die so do we . . ." reminding us that 70 percent of the Earth's surface is ocean.

When trees are chopped and species are displaced and pollution grows, people are affected. There are other dots to be connected too. Richard, the interpreter, for example, says he has noticed that the winds are stronger in Borneo now (where he has lived his whole life) because of the loss of tree protection due to logging. Animals are pushed into places they shouldn't be, upsetting the ecosystem and posing dangers. In a tongue-in-cheek advertisement by the Hong Kong and Shanghai Banking Corporation (HSBC) for its commitment to help the WWF in Borneo, the situation is made clear:

MAN & TIGER: A DIALOGUE

MAN: *Yesterday you ate one of my cattle. Can you comment on that?*

TIGER: *Yes. It was delicious, thank you. I prefer venison but the deer seem to have vanished with the forest.*

MAN: *I apologize for that. But please my cattle are not for you.*

TIGER: *I don't understand. You let your cattle wander. Dinner practically walked into my mouth!*

MAN: *Well, I can't keep them at the village all the time. They need to graze!*

TIGER: *Have you never heard of a schedule? I usually work the night shift.*

MAN: *It's funny, though, the village over the hill—no one's lost a cow before.*

TIGER: *Oh, that one. I've been there. They house their cattle in these enclosures I couldn't get into. Not that I ever tried, of course. The villagers even cleared the underbrush. There was absolutely nowhere to hide. Two years without a single ambush. It was absolute misery. I had to leave.*

MAN: *Only to move here, of course. Now the enclosures, that's interesting. How high did you say the fences were?*

TIGER: *All I will say is I have no difficulty scaling walls two meters high.*

MAN: *I must admit, that's rather impressive! It's good we had this chat.*

TIGER: *It's been a pleasure. We'll catch up sometime.*

As we turn our forest into farms, tiger habitats grow smaller. Displaced and starving, tigers attack our livestock, or worse, us. In return, we kill them. HSBC and WWF-Malaysia are educating affected farmers on better livestock management and tiger-proofing paddocks.

Dr. Colleen Howell, who performed the research for much of this book, did her dissertation on the importance of rainforest products such as fruits, nuts, vegetables, rattan, bamboo, and medicinal herbs as well as certain wildlife to the subsistence and income needs of Malaysia's indigenous people—the Orang Asli (literally, "original man").

Dr. Howell says:

> Most Orang Asli live in remote regions on state-owned reservations, from which they can be relocated if the government decides to exploit the land for another purpose. Urban sprawl and the acceleration of oil palm cultivation have displaced some Orang Asli settlements, forcing them onto marginal lands where they have no choice but to reestablish a viable way of life in a new region. Finally, Orang Asli groups are impacted when forests are cleared—for timber, to plant a cash crop like oil palm, or to build a dam. An estimated 40 percent of Orang Asli live in or near rainforest areas. Because of a lack of regular wage employment, rainforest products can be crucial for supplementing food and cash needs.

Dr. Howell found that forest products were effective at equalizing income, resulting in less income disparity between the wealthier and poorer households. But when the forest products are taken away from the people, the margin of income disparity becomes wider and wider; some people who were living well above the poverty line sink despairingly below it.

That's an academic way of saying that when you take the forest away from people who largely live off it, you are forcing them to seek out other ways to live—which is what the government of Malaysia wants. Malaysia has created what it calls the Multimedia Super Corridor, or the MSC, to create and attract businesses ranging from real estate to telecommunications and technology, and to increase job growth, stock growth, economic growth—all growth.

The Malaysian government admits that the MSC will corrupt land, but says the economic benefits of jobs and industry will outweigh that and help its people.

This is the tradeoff of industry and big business. It is how lands get lost, species go extinct and we justify it all under the banner of helping more, doing better, becoming "civilized."

A great piece of technology may come out of Malaysia. A new skyscraper. Maybe even jobs for thousands of people who will then prosper and increase their standards of living. Great for humankind, but what about the rest of the species?

The orangutan is the only great ape in Asia and is one of humanity's closest living relatives. Its name is translated literally as "jungle man." They mostly live here in Borneo, but the future of the orangutan looks bleak. They are not only victims of illegal poaching and face a diminishing forest habitat, but are often killed for "trespassing" onto plantations where they have been found to be eating palm seedlings. Fewer than 60,000 orangutans remain on the islands of Borneo and Sumatra, and roughly 5,000 disappear per year. The United Nations predicts that orangutans could be extinct in the wild within a decade. It says the population has been cut in half over the last ten years alone and is down from about 230,000 just a hundred years ago.

Let me repeat: the orangutan is one of our closest living relatives.

—   —   —

WHEN I FIRST step into the Borneo jungle I don't think it's such a big deal. We park our car on the side of the road and meet the Iban tribesmen guides. They are dressed in unremarkable long-sleeve shirts, three-quarter-length pants, baseball caps, and flat lace-ups, like those 1980s jazz shoes with no socks. The only hints that they are tribesmen are the machetes attached by rope to their waists and the woven basket cylinders they wear on their backs like knapsacks. No one speaks a drop of English except Richard, the translator.

We make our foray into the jungle, in single file. I take up the middle spot behind Richard. We march fast—real fast. The tribesmen don't miss a step even though there are small creeks across which we must balance on logs, and twists and tight climbs uphill and downhill over the rooted and rocky trails. All this in just the first fifteen minutes of our jungle hike. This is not going to be the piece-of-cake Sunday hike that I take each week at home in the Santa Monica Mountains. And it's sweltering hot and humid.

I focus on my footsteps and get into the groove of the pace. There is a calming feeling to being out in the middle of the jungle. Here we join the rest of the species, without industry or the products of it. It's just our selves walking in step with all the other living things in the world.

The jungle light falls obliquely through the dense tree cover, filtering its way to some plants before reaching the ground. At other times it merely rests on a leaf or a branch or a vine. It's a light that peeks in at you but never fully reveals itself or the sky above. It creates a sense of mystery, as if it might completely disappear—or open up wide and big and bright at any moment. There is a tantalizing notion that is the theme of the jungle: the unknown; and it makes me feel alive.

My only regret is my position in line: Richard in front of me smokes and farts. I need to get out from his tailwind so I inquire about some trees I notice that are leaking white fluid. At the bases of these trees are aluminum cans chopped in half, cups, the tops of plastic bottles turned upside down—all filled to various levels with that white stuff.

We stop at my query, and my guides explain what is going on. Early in the morning the Iban go into the jungle and gash certain rubber trees. They take leaves and point them downward, one on top of another from the gash to the container they place at the tree's base. Out of the gash flows latex. At midday the Iban come and collect the containers.

I'm shown a small camp where the next part of the operation begins. There, spread out, is a colored towel about three feet by five feet and a bottle of pure alcohol. They mix the latex with the alcohol and spread it out onto the towel smoothly and evenly. This way the latex takes the form and shape of the towel. They let the latex dry and then peel it off. Voila, a salable sheet of latex. Richard says they get five ringgit per kilo for the latex. That's about a dollar and a quarter. Richard says this as if it's a really good price.

We trudge on, this time with Richard behind me, and head deeper into the jungle. All the while, however, I can't help thinking that I'm trekking through a forest of condoms.

As I am traveling in Borneo a cyclone has torn through Bangladesh killing three thousand people. The storm had been predicted, but it whipped up at a faster rate than had been expected. A cyclone is a generic meteorological term for inward spiraling winds. They rotate counter clockwise in the Northern Hemisphere and clockwise in the Southern Hemisphere. They are strong storms that appear quickly like tornadoes. In the United States, we know them as hurricanes.

We can evacuate places, build levies, inform people to take more environmentally sound steps where they live, but without help from the natural environs, we are in real danger.

Take Bangladesh. What kept the devastation from being worse than it was were the mangroves. The Sundarbans National Park is a vast mangrove forest, listed as a World Heritage Site by UNESCO. It is a natural barrier that stands between much of southern Bangladesh and the Bay of Bengal and offers protection for the low-lying country from the ever encroaching Indian Ocean. Mangroves stave floods, waves, and erosion. In India Debi's Conservation Action Trust is trying to help save the mangroves along the coasts of Mumbai. In Borneo the WWF is campaigning for mangrove preservation as well.

Mangroves, trees, and shrubs that grow in coastal habitats in the tropics and subtropics are natural protectors and we should be carefully minding them, but instead, we are destroying them. Dr. Howell says:

When rainforests are cleared, localized desertification can result, as land that was once protected by the misty forest canopy becomes baked and crusted-over in the hot tropical sun. The absence of vegetation increases the risk of flooding, which not only threatens human settlements, but removes the nutrient-rich topsoil of the land, making it unsuitable for plant growth.

In short, deforestation dampens our natural protectors—mangroves.

The creek crossings we made early on the trail were nothing compared to the river crossings we had to make now. The Iban pick a spot across the river and walk right to it. Seems easy enough. But it isn't so easy to cross a moving river while managing rocks and mud hoping not to fall in. It is only when I get to the other side of the river and look down at my wet and muddy trail-running sneakers and socks that I realize the genius behind those crazy-looking dance shoes the Iban are wearing: they are made of plastic and can be washed off easily.

We come upon a teenage boy wielding a chainsaw. He is the one making all that racket in the jungle. He's hacking away at a giant felled tree. Several yards behind him a bulldozer is clearing the land down to dirt and raking it with its large steel teeth. This is the logging crew. They work in small teams like this clearing acres of land and hauling timber to logging "ponds," where the stacks are taken away by either helicopter or truck. There aren't vast logging operations on hundreds of acres in Malaysia any-more. There isn't enough forest left for those types of large-scale operations. Instead there are camps like the one I'm in spread throughout the jungle.

The bulldozer lurches toward a large tree in its way. It raises its bucket and smashes the tree. The teenager with the chainsaw runs to it as if it were fallen prey and saws off the root and the branches.

Moving past this camp to the next station gets difficult. When the trees are felled, the unearthed roots loosen the soil. When it rains, as it does here every day in the rainforest, mud sops the paths. A thick mud is waist-deep in some areas. Here we slog on.

At the next station another teenage boy and his wife are working. A blue plastic tarp covers four pieces of wood and a platform. This is their home. A propane cooktop stove and some plastic dishes. A dozen fish skeletons in a bucket. Some fruits and vegetables. This is their kitchen. The wood platform their bed. On the ground is a battery. There is one lightbulb and some wire. Everything is exposed to the elements.

They've made this their home for the past four months and are planning on staying another six. They are illegal workers from across the border in Indonesia. You can see the mountains that serve as the border miles away.

For every tree they chop, the couple says they receive three ringgits. That's less than a dollar. For the past few months they estimate they have earned about $120. Meanwhile, the owners get market prices, which (depending on the type and size of tree) ranges from several hundred to a couple of thousand dollars. And even then there is a scam. The owners tell the workers that up to one-third of the trees are "no good," so they don't pay them for those. Of course, they take the trees anyway.

Every tree of this size represents about 700,000 sheets of paper. Considering we pay about $3 per ream at our local stationery store, those trees turn into $4,200. Then there's timber prices. A solid piece of 1' by 8' mahogany—the most exported wood from Malaysia—goes for about $80. It's used for furniture, doors, floors, decks, even roof shingles.

In a few months this area will be cleared and the workers will be gone. The bulldozers will have raked the land and the wet season will be over. It will be time to plant palm.

Palm oil is made from the fruit of the oil palm tree, *Elaeis guineensis*. These trees look, well, like palm trees except that the type of fruit that grows on them is special. Two types of oil are produced from that fruit: the palm oil itself from the skin of the fruit and palm kernel oil from the seed inside.

When the fruit is picked and sent to the refinery, the fruits and seeds are separated. The oil from each is then refined differently, using different chemicals and processes. When it's liquefied, the oil is bottled and shipped around the world on tankers.

Pure palm oil is reddish because it has a high content of carotene. It is semisolid at room temperature and is usually blended with other oils for frying or baking. Potato chips, french fries, and doughnuts are all typically fried with palm oil or some mix of it.

Palm kernel oil, meanwhile, is used in margarine, confections, coffee whitener, or as a substitute for cocoa butter.

Although the palm oil industry boasts the benefits of palm oil—cheap, long lasting, cholesterol free, zero transfats—critics cite the negative effects of palm oil on cholesterol levels themselves. Indeed, the World Health Organization states in a report that palm oil increases the risk of developing cardiovascular diseases. A quick look at the fried food it helps make—potato chips, french fries, doughnuts—and you can see why. These aren't the types of foods found on most nutritionists' recommended lists.

It isn't easy to check product labels for palm oil as an ingredient either. Because palm oil is mixed with other types of oils, it's often generically labeled as vegetable oil. There are six types of palm products: palm oil, palm olein, palm stearin, palm kernel oil, palm kernel olein, and palm kernel stearin. And the palm oil used in frying doesn't make the labels.

According to Payne, the life cycle of a palm oil plantation is about twenty-five years. Many land leases are for thirty-five years, which means after 1.5 cycles of oil palm the land is abandoned. As many of the oil palm land deals were agreed on in the 1970s, leases are coming up now.

The WWF is trying to get plantation owners to embrace sustainable harvesting techniques so that they see the value in extending their leases rather than just abandoning the land. WWF is also trying to create a labeling system for palm oil products made from sustainable plantations.

With an oil palm sustainability label on products, consumers could make informed choices about the products they eat and/or buy at the supermarket. It's these types of solutions that will make the difference.

As it stands, it's hard to figure which palm oil is made from a sustainable plantation. You'd have to research the product and its manufacturer; figure out where they get their palm oil supply; research the supplier; and then do due diligence on its practices to find out whether they are sustainable farmers. That's quite an undertaking before brushing your teeth in the morning or doing the dishes.

WE LEAVE THE logging pond before other workers arrive and our chances of running into someone who won't be so happy with our "tour" increase. To avoid backtracking through the logging farm again, we take an unmarked route through the jungle. The Iban unsheathe their machetes and go to work making a pathway toward the river. From there we can find a marked path out, I'm told.

It's getting late in the day. We are all tired. At the river's edge we all stop and look at the current—it's running stronger than earlier in the day. Here, the river is wider too.

The Iban raise their machetes over their heads and make their way across. I'm next. On the opposite riverbank the current knocks me off balance. I reach up for something to hold on to and find it: a tree branch. It rights me. Unfortunately the tree branch is laden with long thorns. Six of them pierce my hand and I have to pull them out. I extract five but the sixth one is deep. The Iban guides come over and try to help. One actually gets ready to slice open my hand with his machete—no shit—until I shake my head and Richard advises him against it. Then one of them goes off into the jungle and comes back with a leaf. He breaks it open and a gooey salve appears. He runs it on my puncture wound. It's a natural form of antiseptic, Richard explains. It's common here. Working the salve in, I'm able to pull the thorn out and we get back to the business of finding a pathway out. It doesn't take the Iban long to find our way.

At the jungle's edge we run into a group of locals. They are getting ready for a cockfight.

"Wanna go?" Richard asks. I decline, and we leave the Iban guides there to watch.

By the time we get back to the hotel it's night. The rain stops around dusk during this time of year and the moisture lingers until the sun fully goes down. The wind usually follows in the darkness, winding its way through the jungle and across the river. I am at the river's bend to greet it.

The lights flicker along the riverbank. It's time to leave Borneo. Packing, I count the number of things I have with me: twenty-five articles of clothing, my laptop and electronic accessories, as well as a toothbrush. I have been living with these things for a month now. I don't miss all that I have back home. A figure that sticks in my brain is the number of items the typical US household has: ten thousand.

In the United States we shop until we drop, but it takes a lot of energy to manufacture all that stuff. Moreover, it takes a lot of fuel to transport it—and us—to stores. Energy for transportation and manufacturing causes the most pollution.

Consumption is how we affect the world in ways that we often don't realize. Getting the palm oil in our products, for example, or the lumber to build our homes pushes around our animal neighbors and destroys our plants and places of natural wonder.

I take in a long breath of fresh air. This is what the rainforest is meant to produce—oxygen. It's what ties us to so many of the other species on the planet. We all breathe the same air.

There are about 50,000 vertebrate animals like us humans on the planet. More than 1.6 million species have been identified on Earth. Yet, some estimate that as many as 100 million species exist. We simply do not know our place in and among them all. The award-winning biologist E. O. Wilson calls all species "the tree of life." And we humans certainly don't want to be responsible for chopping it down. "To the extent that we banish the rest of life, we will impoverish our own species for all time," Wilson says.

Keeping our air clean, our forests dense, and our land fertile will keep us rich in natural capital.

The place that may be bankrupting us the most—the core of the pollution problem here on Earth—is where I head next.

FOUR

What Are We Doing?

Linfen City, China

Darkness usually falls from the sky, but here it comes from below.

I am flying from Ghanzou, near Hong Kong, to Taiyuan, a city 125 miles north of Linfen, which for years has been considered the dirtiest place on Earth. Pollution is so high in Linfen that just breathing the air, health officials say, is like smoking a pack of cigarettes a day.

About an hour or so into the flight I glance out the window when we are flying over Linfen: a thick brown and black smog covers the horizon. I didn't expect this—it's so prominent that it's almost unbelievable. The gray and black depths below me show nothing but a soupy haze. We are flying through that mass of pollution, the worst pollution in the world.

This smog is truly shocking. It is like cloud cover with no end in sight. At the skyline a rainbow forms: blue, yellow, orange, red/brown, and then gray/black. These are not the usual colors of the rainbow. Then again, this is a most unusual place; the air is filled with our industrial waste—our gift to the atmosphere.

When I land in Taiyuan the smog is lighter. You can still see it; it is sort of like a dark and cloudy day when it's about to snow. And if you saw it, you might take note of it but you wouldn't be particularly alarmed. You wouldn't think that this part of the world has the worst air quality of any place anywhere. Nor would you think that this is the most polluting place on the planet. But I soon find out how bad the pollution gets.

The reason Linfen and the province in which it is located are so polluted is coal plants. There are forests of coal plants here. A new one gets built every four days. Coal is cheap to burn and easy to derive power from. And these days, China needs a lot of power.

China is growing at a mighty pace. It is the world's fourth-largest country and with more than 1.3 billion people, the most populated. Put another way, one out of every five people on the planet lives in China.

Coal plants supply electricity to the nearby cities, towns, and villages, but more so to the industrial plants that manufacture products—products, of course, that are shipped all around the world.

Manufacturing is the real culprit in creating air pollution. With exports on pace to break one trillion dollars, China is leading the pack in terms of economic growth in the world market. At the same time, China has surpassed the United States when it comes to carbon emissions and is now the world's leading polluter on a total tonnage basis.

Leaving Taiyuan for Linfen, my translator and I board a coach bus. The trip will take approximately four hours as we head south along the famous Silk Route. Don't think this is scenic or leisurely. My knees are up to my chin in my steel seat. The bus is packed and up front a "Hong Kong" movie is playing on a TV monitor, the speaker blaring directly over my head. Hong Kong movies are what the Chinese call kung fu movies. This one stars Jackie Chan. It's as nonsensical, I'm told, as his English-language films. My personal problem is with volume, not film quality. High-pitched sing-song screams along with gunfire make me want to get up and strike a few blows at the volume controller. The driver is also incessantly beeping the horn.

Along the roadside there isn't much to see but industrial plants and wheat fields. It's drab. While the Silk Route may bring images of shiny, expensive material or thoughts of grace and refinement, this journey is anything but charming.

Just a few miles outside Taiyuan the air becomes smoggier. I can see maybe a quarter of a mile ahead. Another hour goes by. I can see about an eighth of a mile away. Closing in on Linfen, it gets worse. There is not a leaf on a tree or a green blade of grass. Smokestacks standing thin and tall scatter alongside the highway. Trucks—big flatbeds—cruise by on both sides, convoy after convoy of them. I assume they are filled with coal but I can't be sure. It's a good bet though. Coal is the industry here.

The Shanxi Museum in Taiyuan has an exhibit that traces the history of coal mining in the area back for centuries. The modern-day result is Orwellian. All the cars have layers of soot on them. People along the side of the road wear surgical masks. Heavy machinery and the barriers to industrial plants—fences, walls, gates—haunt as we pass them by. The sun, whose circle was distinguishable back in Taiyuan, has disappeared completely behind the haze.

This—no one would have to tell you—is the journey to the most sullied place on Earth. If I were a more pretentious writer, I'd quote Dante.

Entering Linfen is like entering another dimension. At dusk, when pollution levels are highest, it's difficult to see more than twenty feet in front of you. As I take a left turn into my hotel driveway, two people wearing surgical masks appear on bicycles as if out of nowhere; they emerge from the smog. It's surreal. Like the environment created in the movie *Blade Runner,* the environs are of another time, another world, or perhaps the not-too-distant future.

To reduce its air pollution to be on par with stable climate levels, China will have to reduce carbon emissions by 80 percent—a seemingly impossible task. But there are alternatives that can be sought—cleaner energy, cleaner coal even. And then there's our participation in this mess.

About 25 percent of the pollution—and 35 percent of carbon dioxide emissions—in China comes from manufacturing goods for export to Western countries. Interestingly, as much as 25 percent of the pollution in Los Angeles comes from the emissions of coal plants—coal plants in China, that is—the winds carry it across the sea.

We are contributing to our own demise and health hazards by the products that we buy and the choices we make. The United States is China's second largest trade partner. Americans buy more than $300 billion worth of goods each year that are made in China. Most of the goods that we import from China are consumer goods, things that we as individuals buy as opposed to industrial goods, which companies purchase. The top things we buy are computers and computer accessories; cell phones and other telecommunications equipment; furniture, appliances, and other household goods; clothing and shoes; toys and sporting goods; and TVs, radios, and other consumer electronics. Just 20 percent of imports from China are industrial supplies and industrial machinery.

The excess pollution these imports create and the excess energy required to transport them are the two main reasons we are urged to buy locally made products. Wal-Mart alone spends $20 billion on goods from China. When Wal-Mart recently required suppliers to ship goods with less packaging, the savings amounted to millions of dollars in energy costs, and incidentally, all the carbon emissions that would have gone along with them. Wal-Mart proudly states that the savings from this new mandate of its suppliers all over the world will stop 667,000 metric tons of carbon dioxide from entering the atmosphere. This is equal to taking 213,000 trucks off the road annually, saving 323,800 tons of coal, and preventing 66.7 million gallons of diesel fuel from being burned. The initiative will also generate $11 billion in savings, just from that 5 percent reduction.

Simple choices clearly add up.

Chinese environmental experts criticize US consumers for buying so many Chinese goods and then complaining about all the pollution that is created from manufacturing. To use Wal-Mart as an example again: thirty thousand Chinese factories supply the store with its goods.

Chinese environmentalist Plato Yip, the vice chairman of the Hong Kong People's Council for Sustainable Development, recently held a roundtable discussion among US businesses, the Chinese government, and environmental groups in Hong Kong to come up with some type of roadmap for fostering economic success while at the same time mitigating pollution.

"Yes, it is absolutely true that US consumers want stuff cheap and for that they look to China," Plato tells me. "CO_2 emissions come along with all those orders, of course."

China's gross domestic product growth is in the double digits. Inflation in China is estimated at between 6 and 8 percent. And, "We know that pollution costs 5.8 percent real growth. So, we have a situation where real pollution causes negative growth," Plato explains.

That isn't good for anybody. And the blame game that tosses the issue of pollution from country to country is building tension between the Chinese government and US businesses. The Chinese province of Guangdong, for example, has threatened to sue US businesses for libel to protect its reputation; the province claims it's being labeled a hazardous polluter.

Not only can we Americans change our consumption patterns, China can also take some simple steps to clean up its act.

Before the 1984 Olympics in Los Angeles, the city persuaded factories to shut down and kept drivers off the freeways. The result, according to the Environmental Protection Agency, was a 12 percent drop in ground-level ozone, or smog, levels. So the strategy worked. In less than a month, LA was able to clean up some of its famous smog. Seoul implemented a similar pollution control plan in 1988 as did Athens in 2004.

In a short period of time, pollution can be reduced.

Beijing is trying its own brand of pollution control for the 2008 Olympics, including shutting down factories in the area, as LA did. But there's a bit more of a problem in Beijing. The Olympic Games are held in August when the wind changes direction. That means pollution can blow in from heavily industrialized provinces. Indeed, 70 percent of Beijing's summer pollution comes from outside the city. As *Wired* magazine reports, "you could shut down the city, close the highways, turn off the power, and still have a seriously bad air day."

The air quality concern, which raised the eyebrows of the international community because of the Olympics, is a reminder that the problem isn't local or so easily fixable; it's nationwide. All of China has to take radical steps to become more environmentally friendly. And we need to pitch in and help. Within the list of the top twenty-five goods the United States imports from China each year, you'll find more than a dozen everyday items:

electronics, clothing, furniture, tableware, shoes, exercise equipment, etc. You may be surprised, however, to find that toys and table games rank third in terms of the dollar value imported: more than $13 billion per year.

Reports of lead-tainted toys from China brought the problem of China's pollution directly into the living rooms and playrooms of thousands (if not millions) of US homes at the end of 2007. Despite the alarm this raised among parents and health officials, it's important to recognize that contaminated toys are not the result of malicious Chinese toy makers trying to ruin Christmas. Contamination has become ubiquitous within China's manufacturing sector, which means that toxins are infused within many of the products made. With the rapidly expanding global marketplace, when dangerous substances become unavoidable in China, they'll become unavoidable on our shores as well. Try as we may, it is very difficult to stay away from the Made in China label, especially because many of the minor ingredients (like spices) found in food come from China, but their country of origin is not required to be revealed. Aside from an outright boycott, we'll see real change occur only when consumers insist that US companies require their manufacturing operations in China to abide by more stringent environmental standards. Currently, many companies have moved operations overseas in order not only to take advantage of cheaper operations costs and labor, but also to avoid the types of regulations that are meant to protect people and the environment at home.

If nothing else, we should look at China as a bad example for ourselves.

THE LINFEN RIVER flows into the Yellow River, one of the most polluted in the world. Local clinics there are seeing growing numbers of cases of bronchitis, pneumonia, and lung cancer. Lead poisoning is also at very high rates in Chinese children in the Shanxi Province where Linfen is located. One resident was quoted in a radio report claiming, "I feel like my throat is very dry, and the stuff coming out of my lungs is black." My own throat became scratchy and sore within two days of staying in Linfen. By the end of a week, I was sick.

In the most polluted parts of Linfen, the death rate for those fifty-five years and older is now 61 per 1,000 each year, more than ten times the normal death rate in China.

Beilu village, on the outskirts of Linfen, is known locally as "the cancer village."

And the ills of Linfen are not too far off from what is happening at home.

More than one in three Americans live in an area with unhealthy air, and in many areas it is getting worse. One out of every three asthma victims is a child. Put another way, 8.5 percent of children suffer from asthma, which is the number one cause of school absences. Air pollution causes other damage to the environment, including forest damage from acid rain and ozone, and eutrophication (overfertilization from nitrogen) of lakes and ponds.

Look at what happened to Lake Tai in China. Toxic cyanobacteria, commonly referred to as pond scum, turned the lake fluorescent green. The stench of decay choked anyone who came within a mile of its shores. At least two million people living amid the canals, rice paddies, and chemical plants around the lake had to stop drinking and cooking with their main source of water. Chemical plants were the source of the problem at the lake: untreated chemicals discharged from the 2,800 factories that dot the lake's shore led to the toxic cyanobacterial bloom.

Wu Lihong, a lake area resident and environmental activist, alerted people to the toxicity of the water as well as the causes.

He was jailed by the Chinese government. As the *New York Times* editorializes,

> Pollution has reached epidemic proportions in China, in part because the ruling Communist Party still treats environmental advocates as bigger threats than the degradation of air, water and soil that prompts them to speak out. Senior officials have tried to address environmental woes mostly through pulling the traditional levers of China's authoritarian system: issuing command quotas on energy efficiency and emissions reduction; punishing corrupt officials who shield polluters; planting billions of trees across the country to hold back deserts and absorb carbon dioxide. But they do not dare to unleash individuals who want to make China cleaner. Grass-roots environmentalists arguably do more to expose abuses than any edict emanating from Beijing. But they face a political climate that varies from lukewarm tolerance to icy suppression. Fixing the environment is, in other words, a political problem.

We can send messages with the things we buy. We can avoid Chinese-manufactured products and foods. We can try to implement changes in our own country to lessen pollution at home. Businesses too can join the sortie on China with environmental mandates and supplier discrimination. But government policy will wage the real power in the battle to lower pollution levels.

At the 2007 United Nations Framework Convention on Climate Change in Bali, a proposal went out among all nation states to reduce their carbon emissions. The meeting was meant to update and tighten the carbon emission targets of the 1997 Kyoto Protocol, which called for limiting carbon emissions to below 1990 levels. The United States is the only industrialized nation in the world yet to ratify the Kyoto pact.

The Kyoto Protocol was based on the United Nations' 1992 Framework Convention on Climate Change. The idea was to create binding commitments by developed countries to reduce their greenhouse gas (GHG) emissions and send a signal strong enough to convince businesses, communities, and individuals to act on climate change. While the Convention *encouraged* developed countries to stabilize GHG emissions, the Protocol *commits* them to doing so. The rules for implementing these commitments were adopted in Morocco (the Marrakech Accords) in 2001.

The targeted cuts are to be met within a five-year time frame between 2008 and 2012, and add up to a total cut in GHG emissions of at least 5 percent against the baseline of 1990.

The Bali agreement calls for developed nations to take on commitments that are "measurable, reportable, and verifiable," and "nationally appropriate." The commitments may or may not include quantified, binding, carbon emission targets.

If that agreement sounds vague and even weaker than the Kyoto Protocol, it is.

Trying to balance the needs of the economy with those of the environment, no one wants to get specific on emissions reductions because they believe it will hamper economic output. It's a global case of "you first" between the world's two biggest polluters, the United States and China, who are at odds over how to keep things fair.

China says limiting its emissions, or the pollution stemming from its industrial sector, puts it in an unfair position in the world marketplace. After all, it says the United States did not have restrictions on its carbon emissions when it grew into an industrial superpower over the past two centuries. Why, then, should China be held back?

It's an understandable question to ask. The problem with it, however, is knowledge. We now know better about the effects of air pollution and carbon emissions than we did when the United States was on its growth spurt to economic superpower status. We cannot go back in time. That would be akin to someone today saying, "Well, all those cigarette makers were allowed to make billions of dollars, so why shouldn't I get the same chance to sell cigarettes in an unhindered way today?"

Taking into account the damage climate change imposes on the economy, China would be far better off to adopt pollution-reducing policies anyway.

The costs of the health effects alone from coal pollution in China are expected to total $39 billion annually by 2020. Air pollution is also driving some extreme weather events, which hinder China's economic growth by between 3 to 6 percent of its gross domestic product, or $70 to $130 billion annually, according to some estimates.

To be sure, efforts are underway to clean up. Solar-energy power plants are being developed, water and energy conservation plans are being talked about, and even recycling programs are beginning. But there is a long, long way to go in China.

It all boils down to how we are willing to pay the price for our wants and needs: environmentally or monetarily.

We can see the price we pay environmentally for our goods by taking a long, hard look at places like Linfen. Or we can pay the price at the cash register instead.

Environmental researchers have looked at an array of ways to lower pollution in China, such as placing emission controls on power plants and industrial smokestacks and upgrading motor-vehicle emissions standards. They have looked at a range of scenarios and associated costs for getting China all the way to zero emissions.

As the journal *Environmental Science & Technology* points out: they found that ambient air quality could be improved by between 5 and 30 percent, depending on the extent of controls adopted, at a cost of only 0.3 to 3 percent of the value of the goods produced. A one-hundred-dollar DVD player, for example, would cost consumers only thirty cents more after emissions controls were put in place.

I'd pay thirty cents more to avoid more air pollution, wouldn't you?

The key for Americans is understanding and awareness. If we knew which products to buy and what additional amounts of money would help the world clean up its act, we'd step up. If we knew the results of our purchases and our habits, I bet we'd even make better decisions.

I read in the *Los Angeles Times* that children suffer most from air pollution because it affects their lung capacity and their growth: "Mounting scientific evidence reveals that exposure to air pollution interferes with the development of children's lungs, reducing their capacity to breathe the air they need."

That gives me pause. What in the world are we doing to our young?

I WAS TAKEN aback by my stay in Linfen. I knew that Linfen's coal plants produce two-thirds of the energy China uses for manufacturing and that these were necessary to keep up with its growing economy, but I didn't expect the pollution to be so bad or apparent.

The haze and the smog were thick like billowing clouds of gas or steam. I had expected the air quality here to be something only a machine could discern. The "overexaggeration" of air pollution I had read about in articles, reports, and papers on China wasn't exaggerated at all. If anything, the situation has been un-derreported and underplayed.

Dump trucks, buses, cars, motorcycles, mopeds, three-wheeled haulers, bicycles, and people on foot all mesh in a hectic jam on streets lined with neon signs in characters alien to me. I loosely calculated about one in every fifty people don surgical masks. That's a lot. Four million people live in Linfen. Picture then eighty thousand people—an entire football stadium audience—staring at you over the white brows of their *kou zhao* as the masks are called in Chinese.

I found out that most people who wear masks own three types: thick, medium, and thin, depending on the pollution level that day. In China, pollution tends to be highest in the winter when coal is burned for domestic heating, and lowest in the summer when the monsoons come and clean the air of most pollutants. Of course, the pollution doesn't "go away." It's transformed into its aqueous state, otherwise known as acid rain.

At the local university in the city, Shanxi Normal University ("normal university" means "teaching university," not "average") I saw one student wearing a fuchsia mask to match her outfit. Fashion, I suppose, has to be considered when medical appara-tuses become a part of everyday wear.

My translator, Katy, is a graduate student at the university. In a year, she figures, she'll have mastered English well enough to get her degree and become a professional translator for a corporation. She introduces me to one of the English teachers at the school, Rod, who arrived just the week before from Queens, New York.

In his housing room—a glorified dorm room—I ask him why of all the places in the world he chose Linfen to travel to and live in as an English teacher. (A lush mountain village in Japan, for example, would have been more along the lines of what I would have chosen.) Linfen is, after all, famous for being the dirtiest city in the world.

"I didn't know until I got here," he says with a thick New York accent. I lap up the subtlety of being able to distinguish where he is from just by the sound of his voice; I haven't spoken with another American in many weeks.

"I landed," he continues, "and then they told me."

"How did you respond?" I ask.

He shrugs. "I don't care. I look at it as undiscovered. People don't come here because of the pollution. It keeps them away. Besides, it's cheap."

That's one way to look at this place, I suppose. The cost of living in Linfen is inexpensive: a full meal is never more than a few dollars; and my hotel, which is considered the nicest in town, is fifty bucks a night. Rod likes the price of beer (a dollar) and the attention Westerners get. When you walk down the street people stop and stare at you. They want photos with you. They gather around and follow. Once in a while, someone says "hello" and giggles.

Poor Katy fields all the inquiries about me. Is he German? Canadian? There are few Americans who pass through here. What's he do? What's he want? I get a little anxious at the last question considering what happened to the activist Wu Lihong at the lake. China doesn't look kindly on journalists either.

The China Daily, the national English-language newspaper, to its credit, has at least one story a day devoted to the environment, from "Rising sea levels take toll on nation" to "Provinces construct green energy plants" to "Waging a tough war against plastic bags." The headlines mirror many of the subjects that are written about in the United States. Actually, all over the world the "green" theme makes headlines. It isn't just an LA–NY phenomenon as much as many would like to think. The world is concerned. It's within governments where words fail to incite action.

There are actually laws on the books in the United States that require utilities to use the cheapest source of energy to protect consumers from rate gouging. A dizzying amount of utility regulation exists on the state and federal levels to keep prices low. State public service commissions, for example, typically ensure that "consumers receive safe and reliable utility service at reasonable rates." Although considerations are now changing to take into account environmental effects, the focus is largely on rate pricing. Coal, being cheap, has been the natural choice, then, for utilities to use. Going forward, there will need to be a balance between lower rates and efficient conservation incentives when regulating, even the National Association of Regulatory Utility Commissioners admits.

It costs about fifty dollars for a ton of coal. Every ton of coal burned generates enough electricity to light four hundred homes for a day. Yet that coal also releases pollution. One ton of coal burned emits about 2.5 tons of carbon dioxide and twenty-seven pounds of sulfur dioxide—a precursor to acid rain—and a fraction of an ounce of mercury. Over the next twenty-five years, coal-fired power plants will expel as much pollution into the air as has been emitted throughout the course of history. This means that in just a quarter of a century, we will release the equivalent of centuries-worth of carbon dioxide and other greenhouse gases into the air. Collectively, the United States, China, and India will over the next five years send 2.7 billion tons of carbon dioxide into the atmosphere each year from burning coal alone.

The Canadian Clean Power Coalition describes how coal power works: Coal from the mine is delivered to the coal hopper, where it is crushed to two inches in size. The coal is processed and delivered by a conveyor belt to the power-generating plant. The coal is then pulverized into a fine powder, mixed with air, and blown into the boiler, or furnace, for combustion. The coal-air mixture ignites instantly in the boiler. Meanwhile, millions of gallons of purified water are pumped through tubes inside the boiler. Intense heat from the burning coal turns the purified water in the boiler tubes into steam, which spins the turbine to create electricity. The turbine is a massive drum with thousands of propeller blades. Once the steam hits the turbine blades, it causes the turbine to spin rapidly. The spinning turbine causes a shaft to turn inside the generator, creating an electric current. Once the electricity is generated, transformers increase the voltage so it can be carried across the transmission lines. Electricity is delivered to substations in cities and towns, and the voltage flowing into the distribution lines is reduced, and then reduced again to distribute electricity to customers. There, in a nutshell, is the simple process of how electricity is created out of coal. It's been done more or less this way since 1882, when the first practical coal-fired electric generating station, developed by Thomas Edison, went into operation in New York City to supply electricity for household lights.

What happens while all that electrical generation is going on is where the environmental problems lie. Again, according the Canadian Clean Power Coalition: burning coal produces carbon dioxide (CO_2), sulfur dioxide (SO_2), and nitrogen oxides (NO_x). These gases are vented from the boiler. Bottom ash, which is made of coarse fragments that fall to the bottom of the boiler, is removed. However, fly ash, which is very light, exits the boiler along with the hot gases. If unfiltered, this ash would float up into the air in the same way that ash from a newspaper does. To prevent vast amounts of flying ash, most coal plants have something called an electrostatic precipitator (a huge air filter) that removes almost all the fly ash (99 percent) before the flue gases are dispersed into the atmosphere.

Fly ash and bottom ash are removed from the plants and hauled to disposal sites or ash lagoons. Sometimes this ash is sold to the cement industry for construction; you can find it in concrete cements.

Just one coal plant can produce half a million tons of ash per year. The world could become completely covered in soot if filters are not utilized.

The United States is often referred to as the Saudi Arabia of coal. There is so much coal here that we can maintain our current usage for a couple of hundred years. We even export it—mostly to Canada, Central and South America, and Europe.

While we hear talk of alternatives to coal in the form of nuclear power, wind power, solar power, and hydroelectric power, it is going to be difficult to replace infrastructures necessary to create the equivalent amount of energy that coal produces.

Clean coal is another story altogether. It's more expensive to build a clean coal plant, about $1 billion versus $800 million for an average plant, but legislators and industry players are getting behind the technology and these plants are gaining momentum.

Clean coal means converting coal into gas before it's burned to create steam. It works off an "integrated gasification combined cycle," or IGCC. Because IGCC plants don't actually burn the coal itself—they convert it into gas and burn the gas—they can selectively pull out the resulting emissions, including carbon dioxide, and convert them into some sort of solid form (carbon dioxide can be turned into baking soda, for example) and bury them rather than release them into the air. That's right, burying carbon and pollution is possible. And, scientists say, it's not harmful to the Earth because the places in which the gases could be buried are the places from which they came: coal mines. According to the United Nation's Intergovernmental Panel on Climate Change, "2 trillion tons of carbon dioxide could be stored in old coal mines, abandoned oil and gas fields, and in various other geologic formations around the world." That's enough storage for 320 years of carbon emissions from coal plants.

China and other nations are also experimenting with clean coal and other alternatives. Some are burning garbage to generate power. Researchers have found that garbage shrinks 90 percent in volume and 80 percent in weight after being incinerated. That said, once incinerated, two tons of garbage can produce the same amount of energy as one ton of coal. In addition, the residue can be used as environmentally friendly construction material—just like fly ash. And there is another benefit: reducing landfill waste. One garbage-burning plant in China incinerates more than 2 million pounds of waste a day, and generates 130 million kilo-watt hours of electricity a year, enough for 100,000 families.

Certainly the time has come for alternative energies to be developed. Almost everyone realizes this. The question now is who is going to suffer the cost, who is going to sacrifice for the greater good, because that is what we are talking about at the end of the day.

We can easily afford to change the world's energy sources around. According to Daniel Schrag, a Harvard University geo-chemist who studies both ancient climate and carbon sequestra-tion, "Right now we put about 2.5 billion tons of carbon from coal burning into the atmosphere each year. An order-of-magnitude estimate for capture and storage is something like $100 a ton. That 2.5 billion tons is only $250 billion a year—about half of a percent of global GDP. It's a lot of money—it requires political will—but it's not a ridiculous amount of money." Indeed, he says, "solving the climate problem altogether—completely rebuilding our energy infrastructure—is something like a $400 billion-a-year program. The U.S. share is maybe $100 billion. That's not much compared with defense outlays." We spend more than four times that amount on our military budget alone per year, to put those numbers in context.

As consumers, we can do our share too, by using less energy or being smart about when we do use it. People on average use the most amounts of energy between the hours of noon and 6:00 p.m.

daily, the most in summertime. We can lessen the strain on energy grids and try to use energy at night when usages, and rates, by the way, are lowest. Running dishwashers or charging cell phones, can all be done in the wee hours just as easily as in the middle of the day. Those are simple shifts in habits we can all make.

In as much time as it takes me to get to the gym in Los Angeles—fifteen minutes—I get to a coal plant in Linfen; coal plants are that close to major cities here. The plant is walled and sits next to an open field. Next to the field is a pit, a football field–sized ditch where coal is dumped and loaded on its way to getting burned.

I head down a dirt side road and make my way around to an opening by the smokestacks. The wall here is low enough for me to climb and stand on top of. Inside the coal plant grounds are mazes of pipes, machinery, steel girders, and brick buildings. The facility is humming along, almost peacefully as smoke spews. It stinks.

I see him coming from way across the yard: a German shepherd. His tail is down, and he is very focused on getting to where I am. As a dog lover, my first instinct is always a welcome greeting. But this dog is not the type I want to stick my hand out to for him to sniff.

The bark doesn't come until he is at the wall below me. He goes vertical trying to scratch his way up. He breaks left, then right, circling back; he wants to get me. Useless, he stays below me, barking away. No one is around to hear him.

I jump down on the other side from him. Another coal plant is just down the road. Nothing special is going on there either. Truck after truck comes hauling coal, then leaves empty. These plants are just cogs in the wheel of mass production. Some of the plants operate illegally. Some are cleaner than others. But the supply chain to the world is in place, and little will stop it—unless we create demand elsewhere or rise up for better air quality standards worldwide.

Everyone in Linfen talks about the pollution and how dirty the city is. There is a sense of helplessness about it, sort of like that of a futile guard dog, or the feeling many people in the United States have about the government's lack of global warming policy.

When I take off from the airport, I'm glad to leave China. It's been a grimy experience. I realize, however, that a big part of China is coming with me. The sneakers I am wearing, my belt, the pen I am writing with, my toothbrush, are all made in China. Even the pollution created by making these things follows me out of the country.

The "transport and dispersion" of pollution is complex. Many factors influence the way pollution is spread, including wind and atmospheric stability, as well the local terrain. Still, air moves from areas of high pressure to areas of low pressure, and what is known as the "Coriolis effect" will cause it to move to the east in the Northern Hemisphere, and to the west in the Southern Hemisphere; I'm headed east, so pollution is tagging along.

Pollution is one thing. But its effects on climate are a whole other story, a whole other chapter. Carbon emissions and pollution are steps on the path toward more global warming. We'll see what that brings in the next chapter.

FIVE

Distant Consequences

Shishmaref Village, Alaska

The wave rolls in to shore and washes up. It recedes inches before meeting the jagged edge of ice protruding from the permafrost. The saltwater leaves a glistening coat of foam on the sand. The winter sun's oblique light is beginning to set, heightening the shimmer. It shouldn't be this way. The sea should be frozen by now. The eight feet or so between where I am standing and the shelf of land that forms above should be thick with ice. The waves crashing to shore are loud reminders of a new era that has reached this distant beach on the Arctic frontier.

The august days before winter arrives in Shishmaref Village is the time of freeze. It is the time when temperatures begin their descent to minus degrees. It is when hibernation begins, when food is stored for winter, by animal and man. It is also a time when ice is needed to maintain the environmental accord reached thousands of years ago between nature and its inhabitants.

But temperature isn't agreeing this year. Ice refuses to form on the Chukchi Sea, where I am stationed, within walking distance from the Arctic Circle, observing the realities that pit a subsistence existence against the throes of climate change.

Ancestors of the Inupiat Eskimo villagers who live here date back four thousand years, according to archeological findings in the area. They were nomadic, or lived in different camps, until about one hundred years ago when a school was built on this tiny spit of land that sits five miles from the Alaskan mainland, 126 miles north of Nome. Nothing but wilderness extends for miles on nearby land. And nothing but open sea lays offshore.

About 550 people live in Shishmaref. There is a school, a trading post, a native store that sells pelts and local foods, and a few other community buildings. Ramshackle homes are spread out on the three miles from its north point to its south. Just a few years ago, there used to be more homes, until a stretch of them fell into the sea. People then realized something unusual was happening. The climate was changing and wreaking havoc on their existence. They just didn't know how much, or for how long. So they asked for government tests to figure out the extent of the soil erosion on their land. The Army Corps of Engineers estimates that Shishmaref is losing about ten feet of seafront land per year due to erosion. The Corps gives the village until 2015 before it is entirely uninhabitable. It has formally recommended that the village evacuate.

Shishmaref is now widely acknowledged as the first community in the world to succumb to climate change. The problem is, people don't want to move. Or if they must, they want to move as a community. For many, all they know is here: the people, the land, the sea. It is how they survive, and how they have survived, but they now live in a state of hope and fear. For the first time a village elder fell through the ice. A teenager died this year while hunting when the ice he was used to treading gave way. These events were heretofore unheard of because navigating the ice is second nature to these natives. It's like walking to the end of your driveway to get your newspaper in the morning and then one day falling into a hole in the ground. The familiar becomes unfamiliar all of a sudden.

When my charter plane lands in Shishmaref, I notice how narrow the spit of land is, just a quarter mile from lagoon to sea. In the few minutes it takes to get from the landing strip to "town," I can see to my left remnants of homes, where the sea begins to stretch across the Bering Strait to Russia. To my right, on the lagoon side, I see other homes, about the size of trailers, propped up on raised foundations. This isn't a village of igloos and pitched tents. It is a village melded into the twenty-first century by relatively little contact with the outside world. The structures are simply designed: one- and a few two-story buildings. I noticed only three colors on any of them: blue, rust, and green. They are haphazardly designed with ill-fitting and worn wood and timber. Dirt paths make their way in loops around the village, from the airport and trash and sewage lagoon on one side to the—for lack of a better description—hunting refinery on the other.

The hunting refinery is where seals are skinned and hung to dry, as well as where guts and oil are tanked and stored. It's also where the sled dogs are chained up and the wind whistles with howls and barks. Along each side of the island are open-hull boats, moored or beached. Rifle blasts are regularly heard in the distance.

Only a few buildings have running water. Pretty much every one uses the "honey bucket" system as their toilets.

Attires are melded too. A homemade seal pelt cap, maybe even some "mukluk" slipper boots. But in between are jeans, work shirts, parkas, and other clothes that could be straight out of an old L.L. Bean catalog. This is the modern-day Eskimo who washes down seal meat with a can of Coca-Cola, who drives an ATV to his or her dog sled, who practices traditional village Eskimo dances, but walks around with an iPod.

Shishmaref culturally, morally, and geographically is at a crossroads. It is a fascinating example of how an ancient culture survives today, desperate to hold on to its values and traditions, yet faced with the realities the world presents.

I'm bunking at the Shishmaref Emergency Station, which means an old couch or the floor. But at least it has a floor, and walls. For seventy-five bucks the whole place is mine, at night anyway. During the day people work there and villagers come and go. At S.E.S., they track the weather. I ask Sara, who is working at the station when I get there, how the weather, in fact, is looking. She looks out the window and says, "good."

Uh, huh. "Snow?"

She looks out the window again. "No."

"Cold?" I ask, desperate for something she couldn't exactly look for.

"I guess," she responds.

I learn quickly that the villagers aren't big on expansive descriptions or explanations. "Yes." "No." "Good." "Fine." These are common responses. And time. Well, time isn't strictly adhered to. "Meet you at 1:30 then?" "Yes, 1:30." One thirty, two thirty, three go by. "What happened?" "Oh . . . I got busy."

Needless to say, even though this village is falling victim to global warming, it isn't warm—and I mean that temperature-wise, not from an interpersonal perspective. The weather report I checked before I arrived said the temperature was 22 degrees, but with windchill it feels like 12. In winter the temperature in Shishmaref dips to minus 30. So when I am there, it's almost balmy to the locals.

While I scope out the village, the village scopes me out. Replete in North Face attire, I don't particularly blend in. Still, everyone who passes me by waves or says hello. It's a friendly place. They welcome visitors. That makes it easy to track down the city mayor once I ask around for him. People seem eager to help.

I find Mayor Stanley Tocktoo in the carving room at the school. He teaches ivory and wood carving as part of a cultural program for the youth. Along with the carvings come stories, history, and tradition.

Stanley is forty-seven and sports a moustache and glasses. He is slight, but is known as a fine hunter. He recently shot a polar bear that wandered through town in search of food. "He was a skinny one," he says. And months earlier, he chased out another polar bear who had arrived in town. He shot it several miles away. "He was a big one!" he says. "You don't want a wounded polar bear that big in town, so I chased him first, then killed him." Chasing a polar bear takes guts.

Sitting there on a classroom stool he seems demure, as he slouched over etching a seal out of a small piece of walrus tusk.

Carving is another one of Stanley's skills. Another town story tells of a misguided photographer who got too close to a polar bear while he was snapping pictures. The bear snatched the photographer and chomped down on his foot. People eventually beat back the bear until it dropped the man, but the photographer's tennis shoe was left stuck in the bear's mouth. So Stanley carved a polar bear with a tennis shoe sticking out of his mouth that remains a great source of amusement in the village.

Stories through carvings. Stories through dance. Oral traditions passed down from one generation to the next. These are the prongs of identity the Inupiat Eskimos do not want to lose.

"I wish we could get help to move to where it's safe for our kids so we can still practice our tradition, our values, and our customs as our ancestors did for four thousand years," Stanley says.

There is a safe area, he says, about twelve miles down the coast from the island where the culture and tradition could continue.

It's the spot the people of Shishmaref have voted on to be their new home. But moving hundreds of people and their belongings isn't cheap. In fact, a government assessment of the cost of the move totals $180 million to build new infrastructures, homes, and utilities, which is quite a tidy sum for a village where the average person earns a little more than $10,000 per year, and 16 percent of the population lives below the poverty level.

It's easy to see how important hunting is to these people when you factor in the cost of food; subsistence living is free. To subsist off this land, however, you need ice. It's the only way hunters can make their way out to where the animals are, especially the valuable bearded seals. Hunters sled out, harpoon them, club them, and then shoot them. (Those rifle shots I heard.) They bring the bounty back to shore where the seals are skinned and gutted. Families and extended families then share what their hunters bring home. That could add up to as many as twenty-five or more people who live off the kill from one hunter.

I watch as a hunter pierces a seal head with a sharp tool and then slices it open with a knife. It is crude work, rough and inelegant.

"You don't want to shoot them and kill them first," Tony Weyiouanna Sr., a village elder, says. "You want to take them alive." If a seal is shot dead, it can fall irretrievably beneath the ice. Almost every part of the seal is utilized for something. They discard only the head and little flippers. Otherwise, the fur is used to make clothing or blankets. The meat is eaten. The fat is drained for oil. And even the guts are placed in a container. Seal intestines are waterproof and can be used for kayaks, clothing, or tents.

Around Shishmaref, animal parts are scattered: reindeer antlers here, seal pelts there, bones of some sort strewn next to a sled dog's chain and anchor. This all ties back to the hunt.

"We don't like the food you can get at the grocery store," Stanley tells me. "It doesn't taste right." Not only is the taste of commercially made food repellent, so is commercialization. A big concern of the elders in Shishmaref is the influence of the modern world on their children. "Drugs, and alcohol, and all that—we don't want it here," Stanley says. Shishmaref is a dry village. Dark tendencies are already upon them without the amplifiers of substance. Isolation and short days breed depression, so much so that there is a suicide prevention center. At least in the village, they can take care of one another, and look after those in need. The outside world is full of unknowns and anxiety.

"I went to Anchorage once and I felt so lost and I was so scared because I'm used to belonging to a community where I know family and friends and when I went to Anchorage it was totally different. I didn't even want to stay there for the duration of time that I had there," says twenty-five-year-old Leona Goodhope. Her "duration" in Anchorage? Less than two weeks.

"My house was one of the houses that was relocated from one end of town to the other end of town and that was a whole new world. It really is different, just that little bit—new neighbors, new things. We are still trying to get used to it to this day," Leona says.

The lack of ice affects not only hunting, but food storage as well. When the permafrost melts, the natural cold no longer preserves the meat. "A couple of people in my village contracted botulism caused by the hot weather during the preservation process, which is a rare occurrence in Shishmaref," Tony's wife, Fannie Weyiouanna, told the Alaska Climate Impact Assessment Commission, just a week before I arrived.

Shishmaref gets lots of attention as a case study for climate change. Media come and go; government officials show up and ask questions; religious missionaries come to praise the word of the Lord to the village. It's a little bit of a circus. Yet, I was the only nonlocal there during my stay. What I heard about and saw most was how the weather strains the population in so many ways.

Congestion leads to crowded housing quarters, and crowded housing quarters strain the sewage system. The one thing about Shishmaref is its lack of smell. There is fresh air breezing through all the time from the sea because of the various wind currents in the area. But over by the sewage lagoon, past the airstrip, look out, the locals say: it stinks.

Disease has also risen. "More of our kids get the flu," Stanley says. "And we have cancer for the first time among the old." Who knows whether the cases are related to population congestion, but without a doubt disease can be traced to sewage.

Residents of Shishmaref capture rainwater from their roofs as their freshwater source. In an island culture on the sea, freshwater is scarce. With more and more people living closer and closer together because of the erosion of land, sewage spills from honey buckets are more common on the roads. "And when the ATVs go by they splash mud up on the roofs," Stanley explains. It isn't hard to see how fecal matter gets into the water and in turn creates sickness.

Water is the source of life and death here just like in so many other cultures precariously living on the edge of existence. When it freezes, water creates the elixir for the ecosystem in Shishmaref. But when ice melts, it sets off a domino effect of catastrophe. Melting ice affects the permafrost because it exposes land to the elements.

After poking around town some, I walk to the beach where several homes had fallen into the sea and from which others had been moved clear to the other side of town for protection. In autumn, silent storms punish the coastline with massive waves and gusts of wind. Flooding reaches the center of the village, hence the raised foundations of the homes.

"We get silent storms when the water rises real fast. We get two or three storms a year now, when we used to get just one or two," Stanley explains. The storms are called silent because they approach without warning.

Observable tides don't particularly exist here; this patch of sea is too narrow and too far north for the moon to noticeably affect the waterline. Rather, the water level on this island rises with the winds and currents as well as, some claim, excess water from the Arctic ice caps melting. Add some wind power to the high water level equation, and you have the formula for massive erosion and flooding.

"We can lose fifty feet in one storm," Stanley says. When the permafrost melts, whatever land is there is exposed and collapses.

From where I stood on the beach to the flat land above my head, where a dune might be, several feet of a permafrost cliff were exposed. It's this shelf space that has been eroding as the temperature rises. The dirt crumbles and erosion increases. Imagine slicing a chocolate layer cake in half and then taking a bite out of the top layer of one side. The inner portion of that part of the cake is now exposed and more easily falls.

"It's dangerous," Tony, the village elder, says. "People feel unsafe, especially in the fall due to the eroding beachfront."

Moving is a must. It's just a matter of time.

"I don't want to stay here if we're endangered," says Leona. She says she's eager to move before any one else in the village gets hurt. "I'm actually feeling pretty good about it because this village is getting smaller and there's no place else to put the houses and we're losing land every storm that we have and it's getting to be more and more of a bigger portion of land that is being taken away."

It was about a decade ago that the village elders began to notice a change in the seasons. Their spring hunts, the biggest time of year for them as the seals migrate north and it thaws just enough outside for them to head far out on the ice without freezing to death, began earlier and the temperature rose more quickly than usual. The summers stretched and the winters shortened. In between, the silent storms ate more of the land.

"Then we knew," says Stanley.

The native culture relies more on nature's signs and signals than engineering reports. Just as Sara at SES had checked on the weather for me, a look at the sky holds data for these people that computers can only attempt to capture and process.

Percy Nayokpuk, who runs the local grocery, and who traces his roots back several generations to Eskimos in the area, including the famous Iditarod racer Herbie "Cannonball" Nayokpuk, tells me that the snowbirds didn't change color this year like they were supposed to. So he holds out hope for an early, cold, and long winter. When the snowbirds stay white that means they don't have to change color for camouflage in warmer weather. After a pause, however, he qualifies, reflectively: "There was some color on the tips of their wings."

Climate change and global warming are not lost on the locals here. They speculate as to who or what may be responsible for the knock on their lives.

"It's the automobiles from the lower forty-eight," says Stanley.

"It's industry and pollution," says Tony.

"It's the sun," says Percy.

"I don't know," says Leona. "But I know the winters seem shorter and the summers seem longer. I know that."

Their culture doesn't really employ the concept of blame. "It is just something that we choose not to do—blame," says Stanley. "We just move on. We fix things and move on. And we want to fix this. We want to move."

But who or what is really responsible for the problems Shishmaref is experiencing? Is it natural phenomena? Is it man-made disaster? Are the natives bringing on the problems themselves? Could there be another explanation?

According to a report by the US Army Corps of Engineers to the Environmental Protection Agency this is what's happening on Sarichef Island, which Shishmaref inhabits: "The shoreline at the community is being rapidly eroded by storm waves, possibly because the ice pack has been forming later in the autumn than in the past, allowing more of the force of late season storm energy to reach the shore."

The ice pack is forming later in the autumn because the temperature stays warmer longer. And the temperature is staying warmer longer because of the greenhouse effect in which gases and particles allow heat to linger in the Earth's atmosphere. Of course, the places where increases in temperature, even marginal increases, are felt the most are the coldest places on the planet— the North and South Poles.

All sorts of things contribute to the greenhouse effect, but they can be broken down into two categories: natural processes and those attributable to human activities. In geek speak, human activities are called "anthropogenic" causes. (I remember that term by thinking of anthropology, which is "what we do.") What we do includes creating pollution that leads to the greenhouse effect. We do this by burning fossil fuel for our transportation, incinerating our waste, or using fertilizers for our lawns and gardens, among other things.

Natural gases mix in certain unexplainable ways that heat and cool the Earth. To what exact degree any gas associated with the greenhouse effect is responsible for heating and cooling is unknown. Without greenhouse gases, the Earth would be uninhabitable. The average temperature on the planet would dip below freezing and life as we know it would cease to exist. What is known, however, is that since the Industrial Revolution a couple of hundred years ago, the concentration of these gases has increased, tipping the Earth's balance of temperature in favor of heat. Carbon dioxide alone, remember, has increased in amount by almost 40 percent since we began manufacturing mass goods. It is also one of the biggest components of industrial pollution.

It's easy to get lost in the debate over what causes what and who is responsible for how much heat. We have to chuck the mired controversy out the window and look at the basic science of how global warming occurs. Global warming occurs when too much of the sun's heat gets trapped in the Earth's atmosphere. The sun's heat is supposed to hit the Earth's surface and bounce back out. But certain gases and dark particles such as carbon keep heat trapped.

The National Oceanic & Atmospheric Administration (NOAA) reports that the winter of 2007 was the warmest since it began keeping records in 1880. More heat equals more ice melting, which causes flooding and erosion in places like Shishmaref. Of course, other places are suffering from heat consequence too. Greenland has lost depths of ice. The forty-square-mile Ayles ice shelf has broken off Ellesmere Island in Canada. And there's a new crack in the ice leading from Russia all the way to the North Pole. All these occurrences are testament to temperatures rising.

Some new scientific findings show that gas particles in the air aren't the only dark forms of matter that attract the sun's rays and keep them stored long enough to exacerbate the heat effect. Plain old dirt and soot contribute to global warming as well.

Researchers at the University of California, Irvine, have determined that dirty snow accounts for about 30 percent of the increased Arctic temperatures. Dirty snow is darker than "clean" snow, of course, and the dark surfaces absorb sunlight and cause warming. Bright or white surfaces reflect heat back into space and cause a cooling effect. If you've ever worn black clothing on a summer's day you understand the concept of heat absorption.

"When we inject dirty particles into the atmosphere and they fall onto snow, the net effect is we warm the polar latitudes," says Charlie Zender, associate professor of Earth system science at UCI and co-author of the study. "Dark soot can heat up quickly. It's like placing tiny toaster ovens into the snow pack."

Dirty snow has had a significant impact on climate warming since the Industrial Revolution. In the past two hundred years, the Earth has warmed about 0.8 degree Celsius, researchers say, and dirty snow has caused the Earth's temperature to rise 0.1 to 0.15 degree, or up to 19 percent of the total warming. During that time the Arctic has warmed about 1.6 degrees, with dirty snow causing 0.5 to 1.5 degrees of warming, researchers say. That's a huge amount of climate change attributable to dirt and soot.

Where does all that dirt and soot come from?

It comes from other places around the world. Pollution generated on one side of the world can easily be transported to the other side of the world in a few days, according to NOAA. Sulfur and ash and soot bellow out of smokestacks in China—as we've just seen—and are lofted high into the atmosphere, transported across the Pacific Ocean, and then are deposited gradually on the sea ice and snow of Alaska and the Arctic. Analyses show that South Asia contributes up to 40 percent of the soot that makes its way to Arctic and sub-Arctic regions, while North America, Europe, and Russia each contribute between 10 percent and 15 percent of the soot "up there." Distant sources contribute substantially to Arctic pollution.

Dr. Cathy Cahill, a research scientist at the University of Alaska, in Fairbanks, has examined pollutants, including dirty snow, in the Chukchi Sea area near where Shishmaref is located. She says among the materials that end up there are mercury, sulfur, DDT, and sand from the Gobi Desert in Central Asia, 3,800 miles away.

"NASA has tracked it—you can see it online—and it's this huge brown cloud that makes its way up here and travels over Alaska," Cahill says. She explains that sandstorms whip up dust into the air and they travel along with the prevailing winds heading west (and north). The warmer the world gets, the more arid it may get too, producing more dust and sand. She notes that other material pollutants come from industrial plants and mining-smelting complexes in Russia.

And then there's us.

Each person in the United States emits about ten tons of carbon dioxide into the air per year. When we fly in an airplane, we emit a little more than half a pound of carbon dioxide into the air for every mile we travel. When we drive our cars, we emit about a pound of carbon dioxide into the air for every mile we go. When we flick on the lights, we emit about one and a third pounds of carbon dioxide into the air for every kilowatt-hour we use. One kilowatt-hour of use is like leaving ten hundred-watt bulbs on for an hour. Heating our homes with oil emits the most: more than twenty-two pounds of carbon dioxide per gallon. The average home burns between one-half and one and a half gallons of home heating oil per hour. Burning natural gas, the most common way of home heating, emits 12 pounds of carbon dioxide per therm.

All these things contribute to the problems Shishamerf and places like it are experiencing. Sure, we can see the pollution that comes out of our tailpipes and imagine it traveling in the air to distant places like the Arctic. We can probably do the same with jet exhaust because often we see those white stripes in the sky when planes go by. The things we can't see, such as electrical and heating emissions, are the things we should be projecting the effects of because they affect the climate so much. Coal plants, just like those in China, power more than half of America's electricity grids. And we connect wires to those grids so we can have electricity in our homes. Now imagine the smoke and pollution billowing from those coal plant stacks: your light switches and power outlets control it—the more you use, the more it billows. Oil that is siphoned out of the ground, barreled up, and shipped to your basement as heating fuel is actually fossil fuel, or hydrocarbons. Burn it and you release carbon material into the air, so much so that the Earth's natural processes can absorb only half. That means the other half is left over to travel the globe and contribute to its increased warming. And your thermostat controls those amounts because it determines how much oil you are burning.

When you send your trash to the dump, it also affects the Arctic. Incinerating municipal waste sends harmful materials there via the prevailing air currents, according to a report by the Center for Biology of Natural Systems of Queens College in New York. The report says these toxic pollutants, known as dioxins, are contributing to severe health issues in the polar area. Dioxins have been proven to cause cancer, immune deficiency, and harmful reproductive and developmental effects. Dioxins are a class of chemicals known as persistent organic pollutants, or POPs, which accumulate in the fat of animals, namely fish, seals, and caribou—the diet of subsistence hunting cultures like the villagers in Shishmaref. In turn, when people eat these infected animals, dioxins are transferred. For example, dioxin concentrations in the breast milk of mothers in certain Arctic areas are twice normal levels.

Two-thirds of total dioxins emissions are caused by municipal waste incinerators, medical waste incinerators, cement kilns burning hazardous waste, and backyard trash burning, according to the report. That means our trash tossed and burned "down there" becomes pollution and a health hazard "up here."

The pollution is also leading to twice as many girls as boys being born across much of the Arctic. Such dramatic shifts in gender ratios are a sign of neonatal health problems. Endocrine disrupters are chemicals that alter the way sex hormones operate and the way the body normally functions. These chemicals have been found in Arctic pollution. And health problems aren't restricted to just humans. In a weird occurrence, researchers found female polar bears with both male and female sexual organs. Researchers at the Norwegian Polar Institute now believe the deformity may be due to PCBs and other toxins.

Pollution is transported by convection, or the uplifting of air. Convection describes how heat is transferred by things that are fluid, such as gases in the air. Warm air, as many of us know, rises, and is eventually replaced by cooler air. This is the process of convection. This warm air carries dust (and tag-along pollutants) into higher altitudes in the atmosphere. This dust and pollution is transported across the ocean in the prevailing westerlies—high-altitude southwest winds—and then deposited along the northwestern portion of North America. Hence, dust and soot from China, India, and Southeast Asia are found in Alaskan and Arctic ice.

Pollution doesn't travel just by air; it travels by sea as well. It's a much slower process, but ocean flow is the way certain radioactive particles and most pesticides make their way north.

Adding to the Arctic warming problem is the issue of albedo. Albedo is the percentage of sunlight reflected by a surface. White surfaces (like fresh snow) reflect up to 90 percent of solar radiation. When an area of ice melts and becomes seawater, the albedo drops, and only between 6 and 8 percent of the solar radiation hitting the area is reflected, while the rest is absorbed as heat. Lowering the surface albedo by melting snow and ice in the polar regions leads to further warming, and increased ocean temperatures.

As mentioned, the albedo effect on land isn't as dramatic, but the dirt and soot causes absorption of sunlight, which raises the temperature of the ice or snow, causing it to melt more readily.

Additionally, as soot gets swept up in wind it may absorb solar radiation and warm the air.

Warm air is a problem in the Arctic. Twenty-five years ago, a project began to determine the source of Arctic haze, which is a thick, brown layer of fog that can be seen at high altitudes in the region, most visibly in the springtime. At first, it was thought to come from local air pollution or dust from the Gobi Desert. After ten years, the data collected for the project showed Arctic haze contains high concentrations of human-made gases and aerosols. This pollution flows up to six thousand miles eastward into the Arctic basin, where it stagnates just north of the polar front. On some occasions, Arctic haze covers an area two to four times the size of the United States.

Why should we care about any of this? Moving a few hundred Eskimos off an island in the Arctic won't affect my life. Some weird hermaphrodite bear isn't coming near me anytime soon. So, a little bit of snow melts. A big cloud forms. Whatever.

Well, according to research from the National Science Foundation, the accelerating melt of glaciers and ice caps could add from 4 to 9.5 inches of additional sea-level rise globally by 2100. It points out that this does not include the expansion of warming ocean water, which could potentially double those numbers.

Ten inches or so doesn't seem like a lot, but a one-foot sea-level rise typically causes a shoreline to retreat about a hundred feet. Considering that about 100 million people live within about three feet of sea level, with more than 500 million people living within thirty feet, the potential for large-scale displacement around the world is enormous. It would displace the entire city of Miami, lower Manhattan, and Charleston, and more water could even make its way down the Potomac to Washington, DC.

If you recall, half the population in the United States lives within fifty miles of a coast. Combine higher water levels, with warmer water and strong wind, and the potential for natural disasters increases exponentially. That is exactly what is going on in Shishmaref—fast. Like Mumbai, Shishmaref may be the precursor of things to come for us all. But unlike in Mumbai, the effects are worn now.

In the meantime, we are all indelibly connected to Shishmaref and places like it because of the wind currents, the ocean currents, and the particles of matter we emit into the air by burning our garbage, driving our cars, heating our homes.

Standing in the middle of the village in Shishmaref, I contemplate the forces that have led to the ice melting, the erosion, the flooding, the displacement, disease, food shortage, and hunting obstacles here. I thought about how desperate a polar bear must have been to walk right by where I stood. Hunters and animals forced onto a land that is eroding and being swallowed up by nature itself. What would be the next step in the evolution of the world when the top of the food chain has the floor taken away? Wouldn't we end up like the polar bears searching for something solid and familiar to stand on, the ground beneath our feet, humbled if not confounded by nature?

KEN STENEK IS a science teacher at the school in town. He teaches grades seven through twelve. Ken isn't a native. He moved to Shishmaref nine years ago from Washington State to teach. He has since married into the tribe and has two children. Climate change was never his field of study, but over the past three years he has taught it as part of his student curriculum because, as he says, "they live it."

Ken teaches about carbon molecules and their composition and then puts it into the context of the locale environs. He shows me an article he wrote on climate change and teaching. It's a great piece on teaching kids more than just the minimum, as he says, for their development. It's about connecting the dots. He points out that "We define energy, discuss the different forms of energy, and the transformation of energy from one form into another, etc." Then he asks, "Have the students learned about the conservation of energy? Have they learned a little about how the conservation of energy relates to Earth systems? How does this relate to climate change?" He concludes with a statement that I believe holds the answer to my own question about our place in the world changing. He writes,

I do not believe that climate change is an "adult problem" as much as it is a human problem. And although our youth have not created the problem they are the future and they can effect change presently. It is not our role as educators to spread our political agendas or create panic for our students. It's true that over time humans have played a role in climate change through our release of greenhouse gases and other means. But what can students think of to help decrease our role in climate change? I believe that our youths' imaginations are only limited by our willingness to deny or support their ideas. And to give them a "canvas" to find solutions may be one of the greatest gifts we can give to students as we teach them the concepts of global warming.

About a month before I arrived, Ken had his students gauge the temperature of the seawater in the surrounding area. Using data from the National Weather Service, he says, students examined temperatures over the last fifty years and found that the average temperature in winter months had increased by 6 degrees. "They quickly took that information and realized, 'Okay, our ice is thinning, you know, and maybe something with that warmer temperature is not allowing the ice to freeze as thick as it normally does.'" When students wonder about storms, food spoilage, and dramatic shifts in their lives, their education can allow them to better comprehend the world that is changing around them—and perhaps learn from the past to effect change in the future. Or, at least, feel a little safer by understanding what's going on.

It's a lesson, I believe, for all because we, just like those students, are living with climate change, whether we know it or not. Air pollution, eroding seashores, droughts, forest fires, disease, and freshwater issues can all be connected by what we do and what that does to the environment.

Of course we should care about other people. Too often we don't connect our morality with the practicality of everyday things in our lives. If we knew that turning the lights off would prevent erosion from sending Leona Goodhope's home plunging into the sea, or that by turning the thermostat down one degree in winter and up one degree in summer we'd give a polar bear a fighting chance to survive, or by not letting our cars idle we'd give Tony Weyiouanna a better chance of hunting to feed his family, we'd likely shift our habits. We'd put a face with an action.

There are simple solutions we can take. Recycling reduces the amount of waste that ends up in landfills and gets burned, lessening the amount of toxins that find their way north in the air after incineration. Turning off our car engines when we can, taking public transportation, or carpooling reduces the amount of pollution in the air that comes from our tailpipes and rises up, finding its way to the Arctic. Switching the lights off and unplugging appliances when they are not in use lessens the drain on coal plants and the amount of pollution they emit that rains down on the ice.

Individually, these actions may not seem like a lot but together we can stop the temperature from rising and the ice from melting in the Arctic by cutting our carbon emissions in half. Put differently, that means using half the amount of energy we currently use through alternatives, shifts in habit, or more efficient products, such as compact fluorescent lightbulbs, Energy Star appliances, or hybrid cars.

Think of that brown cloud of sand over Alaska that comes from the Gobi Desert. A grain of sand or a particle of dust seems so small and harmless. But en masse they loom large. So too do our personal actions.

It's all too easy to get lost in the hustle and bustle of life and forget about places like Shishmaref, people like Stanley, and our place in it all. It was Stanley's teenage nephew who fell through the ice and died.

Springtime in the Arctic comes with a change of wind patterns, a switch from prevailing westerly winds to prevailing easterly winds. Like these winds, we have the opportunity to switch our patterns. We can and should make informed decisions based on consequences. Indeed, it's time we faced what they are.

When I asked Stanley why any one should care what happens to Shishmaref, he said, "I know we are way out in the boonies, but we care about you. We are Americans, just like you." He said that after 9/11, American flags went up all over the village.

It isn't all up to us as people, though. The Earth has to cooperate in its own way. In the next chapter we see how we can give nature the chance to help us survive.

SIX

Nature's Oxygen Factory

The Amazon Jungle

I am in the cavity of the Earth's lungs; I am sleeping in a hammock deep in the Amazon jungle. The Amazon is often referred to as the Earth's lungs because in the same way that physical lungs infuse the bloodstream with oxygen and remove carbon dioxide, forests generate oxygen and remove carbon dioxide from the atmosphere.

The Amazon is the world's largest tropical rainforest, and its ability to "breathe" plays a big role in the health of our planet. Standing forests sequester vast amounts of atmospheric carbon as they grow, and the Amazon accounts for about 17 percent of the world's forested area. Locked up in the Amazon rainforest, for example, are some 95 billion tons of carbon—equivalent to eleven years' worth of carbon emissions from human activities. When forests are slashed and burned—as is often the case in tropical countries—the carbon held within their biomass (biological matter) is released back into the atmosphere. As pointed out earlier, deforestation and forest degradation make up the second largest overall contributor to global warming, comprising some 20 percent of human-caused greenhouse gas emissions. At current rates of deforestation, the Amazon is losing almost 5 million acres per year—an area the size of Massachusetts.

NIGHTTIME IN THE jungle isn't quiet. The cicadas are on full throttle. In the distance an owl hoots, a toucan screeches, a macaw lives up to its name and caws, and a frog—it must be the biggest one ever—croaks.

I just finished eating a potato I had buried under the campfire to cook. I drank water from the vine of a fiber tree, picked berries and fruits. For a little sweet, I had licorice fresh out of the trunk.

It's definitely a bit of *Survivor* out here, although I am not here to challenge myself to live off the land. I am here to discover how we are challenging the land we live off of.

In addition to being a giant air filter, the jungle is also nature's market: you can live bountifully off the forest—that is, what's left of it. The Amazon's rate of deforestation is thousands of miles of supposedly protected rainforest per year. Since 2000, more than 60,000 square miles of rainforest have been lost. And many of those acres of forest are lost for good because no new trees have been planted.

You really have to see the forest here to appreciate its beauty. Different shapes and varieties of trees forming leaves and branches that seem to stand proud and tall in their function of configuring the landscape.

The jungle, however, is a precarious thing. All species seem to know their place, whether it's a plant that forms thorns to protect itself, snakes that blaze green colors to blend in, ants that take cover under thick roots in the ground, or the trees themselves that grow to soaring heights to keep from drowning in the wet season. The difference in water level between the wet and dry seasons here is a hard-to-imagine forty-five feet. The Amazon grows, soaks, dries, and punishes—on a major scale.

The famous law of the jungle applies to all.

That law insinuates dangers, and there are many: jaguars, wild boars, venomous snakes, alligators, piranhas, and tarantulas are all around here somewhere. At night these things tend to come to mind.

I constantly hear rustling in the trees—there, no there, then there. Branches crack, leaves crunch. Fire is solace, and I tend to it diligently at night. Here, by the way, is about fifty miles west of Manaus, the capital city of Amazonia. I took a ferry from Manaus to an ecolodge three hours away. After spending a night there, I took a five-hour engine-powered canoe ride deep into the forest. Then I hiked in for a half day before setting up camp by a small creek. I didn't bring any food or water. I'm living off what nature provides.

I'm not alone. I am accompanied by Alan, a jungle survival expert and a former military officer who trained Brazilian army troops how to survive in the wild, and Saru, a native of these parts. Saru is an older man with white hair who knows the land, the rivers, and the maze of wetlands like the back of his hand. He's also a jokester. Walking down a trail I'll hear something to one side then another, for yards. I think we are being tracked, but it's just Saru throwing stones to tease me. He'll navigate the boat through a patch of brush growing out of the water to try to tangle me in it. When a woodpecker pecks in single echoing knocks rather than the rat-tat-tat we're used to hearing, Saru says it's because the wood here is so hard the bird gets a headache every time its beak hits the tree. He mimics the woodpecker action, holding his head as if in pain. Nut.

Saru captained the canoe the entire way, steering us through water-covered forests and down small creeks and estuaries. If anything happens to these guys, I am screwed. We are about twenty miles from the last drip of civilization—a shack somewhere in what was then the middle of nowhere. Alan told me a family lived there.

The first thing I do at camp when we settle on which trees to hang our hammocks, and gather wood for the fire, is to breathe the air in deeply and take notes. If this is it, the core of the Earth's oxygen factory, then I want to be able to remember the air.

Breathing in fully, strictly through my nose, I gulp it down. I don't inhale like I was taught when I lived in Amsterdam, I take smooth breaths. "Can't smell a thing," is my note, which is actually a good thing. Sure, there are the smells of wet leaves, dirt, the scents of trees, but the air itself is unadulterated by anything artificial.

This is quite an unusual feat when the air quality of the Earth continues to decline. Over the past 110 years, global emissions of nitrogen oxides (NO_x), for example—which mainly come from burning fossil fuels—have increased by more than 400 percent, and in the United States alone NO_x emissions have soared by a factor of eight.

Nitrogen oxides are not only responsible for smog's reddish brown color, but they also act as a catalyst with other chemicals to further the formation of ground-level ozone (which causes lung tissue damage and reduced lung capacity), acid rain, toxic airborne chemicals (which can cause premature death), deteriorated water quality, reduced crop yields, and global warming. The World Health Organization estimates that more people die worldwide each year from causes directly attributable to air pollution than the number who die each year from automobile accidents.

The World Bank measures air pollution from metropolitan areas all over the world and finds that, in terms of particulate matter, Venezuela has the best air quality, and China, as we've discovered, has the worst.

So what does the best air quality actually mean? According to AirNow.gov, a government Web site developed to inform and educate the public about pollution, air quality tells you how clean or polluted your air is and what associated health effects might be a concern. In the United States, the EPA calculates the air quality for five major air pollutants regulated by the Clean Air Act: ground-level ozone, particle pollution (also known as particulate matter), carbon monoxide, sulfur dioxide, and nitrogen oxides (NO_x). For each of these pollutants, the EPA has established national air quality standards to protect public health. Ground-level ozone and airborne particles are the two pollutants that pose the greatest threat to human health.

Within the United States, the place with the best air quality is Cheyenne, Wyoming, and the worst is Los Angeles, according to the American Lung Association. California has the disturbing trifecta of the country's worst traffic congestion, largest active landfill, and worst air pollution.

The Earth's natural remedy to air pollution can be found in the jungle: its vast forests and trees effectively scrub pollution from the air. Great, isn't it, that Mother Earth has developed her own healing system for all the carbon dioxide that is produced.

You'd think we'd respect this phenomenon, or maybe even help it along. But we do not. The Amazon is being massacred by deforestation. About 230,000 square miles have been lost to deforestation since 1970. At the current rate, the Amazon will be cut in half over the next twenty years.

To put the Amazon in perspective it's important to note how big it is: you could fit all of western Europe under its canopy.

Driving the deforestation of the Amazon isn't just tree cutting for the profit to be made on the trees. It is tree cutting for farmers and factories.

Soybean production, for example, is big business in Brazil. After the United States, Brazil is the largest producer of soybeans in the world. To grow soybeans, you need land to farm. Trees get in the way of that, so growers for the soy industry chop them down.

Brazil's soybean production has doubled over the past ten years, according to the US Department of Agriculture, and so has the harvested land on which it's grown. Right now, the area harvested for soybeans alone exceeds 50 million acres, which is about the size of the state of Kansas.

Soybeans are the most utilized edible oil on the planet. The second most edible oil, if you recall, is palm oil, harvested in another tropical rainforest, Borneo.

The importance of soy on the Brazilian economy has grown significantly since global warming hit the news, and alternative energy sources are now being sought.

Since fossil fuels—like crude oil—exacerbate global warming because of many of the things discussed in this book, there has been a scramble to seek out alternative energy sources. One of these alternative energy sources is ethanol, which can be made from corn. Because it's more profitable to grow corn than soy these days, many corn farmers in the United States have opted to back off their soy production. This has created an opening for soybean farmers in Brazil to pick up the slack and that is one of the reasons for the increased area of land that is being cleared to farm soy. The Nature Conservancy points out how the soy is then used,

Fast-food outlets throughout Europe, including McDonald's, rely heavily on Brazilian soybeans, which are increasingly harvested from fields that used to be Amazon rainforest. The

European Union bought 10 million tons of soy from Brazil in 2006—about 40 percent of Brazil's soy export crop—soy that is used as animal feed to fatten the cows and chickens that become Big Macs and McNuggets.

Nearly 80 percent of the global soybean harvest is milled into animal feed, according to the Worldwatch Institute. The rest of it gets turned into products like tofu, soymilk, and biofuel. Biofuel is made from vegetable oil such as soy, animal fat, grease, or even algae, rather than petroleum. Biodiesel is a common fuel derived from alternative energy sources such as these as opposed to gasoline, which is derived from petroleum processing. It is also ironic that soy, like corn, is being used as a fuel alternative. In 2005, 92 percent of the biodiesel made in the United States and 57 percent of the biodiesel made in Brazil was derived from soy. As demand grows for alternatives to foreign oil, so too will the market for soybeans.

Indirectly consuming Brazilian soy is one thing, but there is also the issue of direct consumption of Brazilian exports that contribute to deforestation as well.

One of those familiar exports is hamburger meat. The Center for International Forestry Research (CIFOR) in Indonesia authored a report entitled "Hamburger Connection Fuels Amazon Destruction," which says that the number of cattle in the Amazon has more than doubled. David Kaimowitz, CIFOR's director, was quoted as saying, "Cattle ranchers are making mincemeat out of Brazil's Amazon rainforests."

Experts say what is driving the demand for Brazilian beef exports are fears of mad cow disease in other beef-exporting nations. The Food and Agriculture Organization of the United Nations reports that in 2004 Brazil surpassed the United States as the largest meat exporter and has maintained around 25 percent of the world market share since then. Brazilian farmers and ranchers see opportunities in world commodities markets and seize them.

Depending upon whom you ask, and how the cattle are fed and raised, you can get as little as 165 pounds of beef per acre of land to as much as 3,661 pounds. No matter, a lot of land is needed to produce the meat that ends up on our tables. Considering the density of forest in Brazil and parts of the Amazon, hundreds of trees have to be felled per acre. (Some studies have estimated there are more than two hundred trees per acre in the Amazon, whereas an average canopy of trees in the US is about thirty-five per acre.)

Make no mistake: those trees don't go to waste. Logging is still a giant industry in Brazil. Wood products are the third-biggest agricultural export behind meat and soy, according to Brazil's Ministry of Agriculture.

The United States is Brazil's third-largest consumer of tropical lumber and its largest market for softwood lumber. We especially like Brazil's pine and eucalyptus trees. Eucalyptus trees are used to build homes or to make pulpwood and eventually paper. Pine is also a common building material.

Natural rubber used to be a big export of Brazil too, until Malaysia and Indonesia entered the picture, surpassing it in world production. Even though the rubber tree (*Hevea brasiliensis*) is indigenous to Brazil, Brazil actually now imports rubber from Malaysia.

How the multibillion-dollar rubber market was yanked from the hands of Brazil is worth addressing.

At the turn of the twentieth century, Brazil's rubber business was booming and the country had a monopoly on it worldwide. Rubber tree plantation owners and government officials closely guarded their rubber tree plants and seeds. Manaus became a boomtown. "Rubbertappers" flocked to the city. It was a "rubber rush" much like the gold rush period in the United States. The opulent Manaus Opera House, which I visited—and which is curiously placed in the middle of the jungle—is a testament to these former times.

During the rubber rush, an Englishman, Henry Alexander Wickham, committed what was said to be the first act of biopiracy: he stole rubber tree seeds, hid them on his ship between banana leaves, and fled to the British colonies of modern-day Sri Lanka and Malaysia, where he planted his loot.

Well, surprise, surprise. The seeds took to the new soil and flourished. Really flourished. The British Empire became the new rubber king. As the new plantations were designed to be more productive than the natural environs in which rubber trees grew in Brazil, they were able to turn out more supply. Meanwhile, Brazil refused to try to compete. With its prices higher and its production slower than what Southeast Asia growers could provide, Brazil soon lost dominance. Then the rubbertappers left the Amazon, the plantation owners moved on to other products, and the rubber business largely vanquished.

Southeast Asia provides about 90 percent of the natural rubber exports around the world today. So much so, that at the end of 2007, Malaysian officials were invited to Brazil to investigate ways to revive the country's rubber business.

This history is endemic to the conundrum Brazil faces. How do you protect your natural resources, but at the same time remain competitive?

Brazil actually has strict environmental laws that protect and conserve the forest and its products. In fact, it has some of the toughest environmental laws in the world in that landowners in the Amazon are required to keep 80 percent of their land as forest reserves. Corruption, however, allows exploitation and deforestation. The Brazilian government even admits this, stating that a majority of the lumber that is exported from Brazil is from illegal logging.

At the end of 2007, Brazil ordered seven hundred federal police officers to the Amazon River basin to monitor and prevent deforestation. More than 1,700 environmental protection agents, police, and soldiers are already in the region for this purpose.

Yet corruption still reigns.

In an example of how corruption and illegal logging work, Alan, who grew up in the jungle outside Manaus, explains that a foreign company will apply for a permit from the government to fell a hundred trees. "They'll pay the right people and then go in and take a thousand trees. Who's to know and who's to stop them?" he says. The company then moves on to another patch and does the same thing. It doesn't bother to replant or think about sustainability in any way. That would eat into their profits, Alan says, and when the company knows it isn't going back to use the same land—land it only temporarily leases and does not own—there is no incentive for them to be conscientious.

This is how the trickle of deforestation swells. Little by little the forest is depleted.

Many reports and analyses show that the leading cause of deforestation in the Amazon is not due to large, multinational corporations going in and damaging the forest. The leading cause of deforestation is unsustainable agriculture, often by local people farming small plots of land, or raising their cattle.

Locals say that is true, but only because local populations are displaced by major corporations coming in and forcing them to move onto new land to graze or farm. The old land they were on is just passed into the hands of the large companies and therefore isn't counted as "deforested."

In a local indigenous village in the jungle that I visited, there was a school and once a month a medical doctor would come by to visit. I asked how this was possible; who was paying for the services? The village elder with whom I spoke—who had the traditional black line of paint across his cheekbones and nose, and was donning skimpy black Speedo-like shorts—told me that in exchange for some land rights, Petrobras, the giant oil and gas company, provides education and medical funding of this sort.

I get the picture.

Petrobras is one of Brazil's leading companies. It is the eighth largest energy company in the world. It is also looking to sell more fuel to the United States under a new joint venture with a Venezuelan company. Other Brazilian companies are looking to export more sugarcane, another big crop grown in the rainforest, to leap on the new biofuel craze. Sugarcane can be readily turned into ethanol as well.

So as companies seek out more product to feed the world pipeline of demand, many of the people who live in the area get pushed out, and more area is taken up to build factories and production facilities. Operations like these also draw more people looking for work. That means more housing and more land use.

Manaus is a good example of growth in the region. Manaus has grown by 65 percent over the past decade or so to more than 1.5 million people, according to the United Nations Environmental Programme, which reports, "In addition to the urban expansion evident in the area surrounding the city, increased logging and road construction can be observed."

And it can. Canoeing along a branch of the Rio Negro, you are suddenly confronted with a bridge being built. Natives, once cut off from the world by all means except boat, can now hitch rides in cars or trucks: the bridges bring with them roads, connecting people more directly with other villages, towns, and centers of trade. This helps them out economically, but it also tears into large tracts of land that they have been used to cultivating.

At a small village of about fifty people on the banks of a Rio Negro river offshoot there's a market where natives come and sell their goods. Some of the goods remain for sale there in the raised hut—where alongside the skin of a twelve-foot-long Anaconda that was recently caught in the river, you can buy a coffee—and some of it moves via land to bigger towns and cities.

A quick glance at a map of the Manaus area and you can see Route 070 nearby. Routes 319, 310, 174, and smaller roads and streets all wind their way through the jungle too. This is a new phenomenon. It's happening, of course, because factories need to move their goods about. Yet it's also the beginning of urbanization taking hold in a land where many people have never been anyplace where there are more than a hundred people.

I met one family that is perfectly happy living off the land they farm. They are five hours by boat to the nearest town. There are perhaps twenty other people living within as many miles. A small creek runs by their front yard where they can fish for bass or piranha. In the lowlands behind the home they built (which actually has satellite television powered by a generator) there are fruits, nuts, beans, vegetables, and grains for the picking.

The couple, beautiful people in their twenties, have two equally beautiful children, a boy and a girl. Their home is immaculate, and they served me a lunch that was one of the best I have ever had: beans, rice, fish, and fresh fruit juice. They seemed happy and not wanting for the company of others or other comforts.

In a shed out back, the husband shows me how to make bread and cereals. Hens and a rooster run about. He says he hunts the forests for wild boar and other meat the local environs provide. He doesn't farm to sell his produce and then incidentally eat what he's grown. It's the other way around, he farms to eat what he has grown and sells the produce left over. He occasionally makes the journey to a town or a local village to peddle the excess.

Brazil is a large country. It is the fifth largest in the world and the fifth most populous. It has all the things one could need or want in a nation-state: oil, gas, freshwater, timber, and commodities of every sort. Its people can live nicely off these resources. It can even sell off its excess to the world market sustainably, much like the farmer does his produce. Unfortunately for all, it isn't working out that way. There is a race for the immediacy of a financial return now, and a subsequent exploitation of natural resources. The "here and now" approach is winning the day and the "what will be good for later" mind-set is sacrificed at the altar of economic viability.

"The upside of globalization is that the rich get richer, but no one is asking whether it is good for the natives in the jungle," says Dr. Mark Plotkin, president of the Amazon Conservation Team organization and a *Time* magazine Hero for the Planet. He tells me, "If you go up to a native in the middle of the jungle and ask him if he wants money, he'll say yes. If you ask him if he wants to sell land and get rich, he'll say yes too. But that is not an informed decision. If you took him up in a small plane and showed him the deforestation or took him to a river that is being polluted where he can no longer fish, he may have a different answer."

As Plotkin wisely observes, native people are far better off having less money and subsisting off nature than being paid marginally more by a corporation for their land and then being forced into an urban slum.

"There isn't a simple answer," Plotkin says. "We need to create economic incentives for natives so they can buy fishhooks and batteries and at the same time provide them with awareness about what deforestation could mean."

In the United States we've conserved less then 11 percent of our land. About 2 billion acres comprise the continental United States. We live on only about 10 percent of this space—75 percent of the population lives on 3 percent of the land—raise livestock on about a quarter, farm on less than 20 percent, manage forests for timber and paper on 29 percent, and the rest is undeveloped.

Meanwhile Brazil farms on less than half the land as the United States, yet it has 50 percent more potential cropland. In fact, almost all of the land, including the rainforests, can be used for some type of crop.

This is tempting space for a country whose gross domestic product is less than one-tenth that of the United States, and whose per capita income is less than one-fifth of that of the US median income.

Stop at any point in the Amazon jungle and look around. If you're an opportunist, you'll find lots of things that you could kill, chop, or snag to sell and make money. Here's a short list: chewing gum (chicle), printing ink (copal), cashew nuts, coffee, chocolate, insect repellent (camphor), beef (cattle), sugar, vanilla, rope (jute), furniture (bamboo and rattan). There's an assortment of fruits and medicinal herbs to be had, and aloe for lotions. There's even natural Viagra.

I WAKE IN my hammock to a new singsong sound in the distance, sort of like an alarm going off. It's 6:00 a.m. There is light, thank God. I woke every two hours during the night to stoke the fire.

It's shower time: down to the small creek to splash water on my face. There, kneeling on one knee is Saru. He's looking at a bunch of crumpled leaves. "Animal," is all I understand. He points to the tracks leading by our camp to the forest beyond. Alan later tells me that it was likely just a tapir.

I smile. Tapirs are the cutest little guys in the forest. They're large—about seven feet long and three feet high—but pretty harmless. They look like a small elephant with its trunk cut short. They are defenseless except if they turn their pudgy bodies around and try to kick you with their small hind legs. They could bite you, but they are very shy. Tapirs feed on fruits, berries, and leaves.

I tell Alan that I could do with a cup of coffee and something to eat. The eating part is no problem. He walks several feet to a bush and pulls off a handful of red berries. "How do you know," I ask, "that they are safe to eat?"

He gives me a rule to remember: no fur, no milk, no bitter taste—and it's safe to eat. Farther away, a palm tree lies on its side. Alan examines it and then hacks it with a machete. Several strokes later he hands me palm fruit to eat from its trunk. The creek has flowing freshwater. I drink that. On the other side Alan also spots something that he says will make me happy. He comes back with a leaf. I gnaw on it; it's bitter. "Coca," he says.

Coca leaves are stimulants, famously the basis for cocaine, but they're also used to make Coca-Cola, teas, and other things. No coffee, so it'll have to do.

Buzz check: Zip.

Alan spends the day teaching me what to eat, what to look for, and how to find my way out of the jungle. The most difficult lesson is learning how to climb a tree for coconuts.

I take a machete and slice the bark off a tree. (Alan assures me it won't permanently harm the tree.) Fibers splinter. I then twine the fibers into a rope and tie that off into a circle. I put one foot in then the other. I pull my legs apart so the circle becomes oblong around my feet. I am next supposed to jump onto the tree so the center of the circle takes the trunk, and my legs shimmy its side. Then, hop, by hop, I am supposed to push my way up to the top.

I try it three times. On the third I am left speaking like Mickey Mouse, and give up. Saru loves this, cackling until he is out of breath.

Later in the day, I am attacked by a swarm of wasps and, just as in Borneo, pricked by thorns I have to pull out of my hands.

Then comes a swim in the river with pink dolphins. A sunset over the marsh. Caiman (alligator) spotting (you can see their red eyes at night when you shine a light on them). And a night sky filled with shooting stars.

This is splendid. This is unspoiled. Before I decamp, I take in a last long breath. It's difficult to leave the jungle.

Brazil, among other countries, wants credit for all noncommercial surplus that it gives the world. By this it means the global benefits that the Amazon provides: climate regulation, the value of undiscovered or endangered species, and carbon sequestration.

Under a program dubbed REDD, which stands for reduced emissions from deforestation and degradation, there is a proposal to pay countries not to deforest and to give them the same carbon emission credits as countries get in the developed world. The idea was proposed at the United Nations conference on climate change in Bali.

Because much of the biggest forest cover lies in developing world countries, where it's difficult to tell people *not* to clear forestland so they can survive and prosper, the UN introduced this scheme whereby these countries would get paid by big carbon emission producers such as China and the United States to keep their forests alive.

Andrew Mitchell, director of the Global Canopy Programme, writing in the *London Telegraph* does a good job of explaining it:

> So what's up with REDD—reduced emissions from deforestation and degradation, to you and me? The answer is, well, lost in translation. If you think the parish council and a local foot path is tough, imagine trying to get 180 countries to agree to anything as complex as fixing this. Think of a steamy Amazon environment, with green fuzzy

tops and roots in the ground that David Attenborough has convinced us is kind of cute. But Pedro with his chain saw needs cash and the forest is a pretty good ATM. Add thousands of Pedros all armed with chain saws, with no laws anyone can enforce, hell bent on proving they own at least something, even if they have to cut it down and put a cow on it. At least that makes a buck. Sooner or later Pedro sells his land to beef ranchers and they sell on to soya barons, who both sell food to us. It all starts with Pedro and he will carry on cutting trees unless someone will pay him more than he can make from the alternatives. That's the inconvenient truth of food for us. REDD is a proposed UN trading mechanism that could change all that, by paying countries not to deforest.

In the Amazon, Eduardo Braga, the governor of Amazonas, has already been piloting a payment of fifty dollars a month to forest-dwelling families who commit to not logging or burning the forest. REDD would extend that payment system to the rest of the world. It won't be easy to figure out who and how much gets paid because there is still widespread confusion over how carbon credits are valued.

Carbon credits work like this: you take the amount of carbon a country, a business, or even you yourself emit (there are organizations that track these things and even provide carbon calculators online), and then seek out ways to offset all that carbon. Countries that emit large quantities of carbon can buy credits from countries that produce low amounts of carbon to offset their emissions. Same with businesses. Individuals can buy credits in "clean energy" programs such as wind or solar power farms. Or we can all pitch in to preserve large tracts of forests, which are natural carbon offsets.

The jury is still out on how much countries, businesses, and people are willing to pay to offset their emissions, but the rationale makes sense: provide incentives for conservation, which benefits us all.

When I'm out in the jungle surrounded by nature I cannot see the infringement of industrialization. The tree cover is so thick that I literally cannot see the entire forest for the trees.

On a relatively small patch of land, the Amazon seems unspoiled, untouched. It's when I fly over it that the damage becomes apparent. For miles there is dense forest, then abruptly it's clear. The barren land shines through brightly. If it were a checkerboard below me most of the squares would be black. But the middle squares would form an uneven pattern of color. Loggers concentrate their operations on the middle of the Amazon and ship their timber along the vast connected water system past Manaus and the port of Belém out into the Atlantic Ocean, where tankers make their way up the coastlines to North America and Europe, or through the Panama Canal to Asia.

It takes about ten years to grow a tree that is specifically planted for making paper. Yet it doesn't take us very long to use the paper itself. Per capita paper use in the United States is about two pounds per day. That's 1,400 sheets per week, or 73,000 sheets per year: the equivalent of almost nine trees per person.

Where all that paper ends up is the topic for our next chapter.

Meanwhile, the solution to deforestation in the Amazon is complex. It isn't stopping our use of paper, avoiding soybeans, or skipping hamburgers. Those things will only pass the buck to other commodities and cut off economic opportunities for native farmers, ranchers, and businesspeople. The solution lies with creating a sustainable extraction and production system for the $600 billion of natural resources the Amazon holds, and figuring out a way to credit it for sustaining our lives too. If the Amazon rainforest continues to be cleared and burned at current rates, the quality of life around the world will suffer: we'll breathe more toxic air and get sick; sea levels will rise more steeply due to increased global warming, and droughts will occur in regions where poor farmers depend on rain clouds that originate in the Amazon to float across the Atlantic Ocean and water their dying crops growing in the sandy soils of the African desert. In short, depletion of the Amazon will mean life-altering consequences for us all.

The REDD program is a logical place for the green movement to begin acting on more seriously. By 2009, when a new global climate agreement is scheduled to be reached by the UN at a meeting set for Copenhagen, adoption of REDD could help steer the world in a new direction, pretty much by doing nothing more than keeping the old world in tact.

Individually, we can buy sustainably harvested wood and lumber products that are certified by the Forest Stewardship Council (FSC). We can buy fair-trade-certified products like fruit, coffee, chocolate, tea, and flowers, which ensures that farmers manage their lands sustainably and are paid fair wages for their work. Local peoples who are paid good wages are much less likely to face the economic desperation that often motivates them to engage in deforestation for extra income.

Or we can seek out carbon offset programs that pay locals to conserve tracts of land. You can also read this to mean paying for the air we breathe. It may just come to that.

SEVEN

The Reality of Our Actions

The Fresh Kills Landfill, New York

B elow me are ships and tankers arriving from all over the world. They make their way through New York Harbor, where the Statue of Liberty stands tall. Even she is an import from France.

I am driving on the Verrazano-Narrows Bridge, the longest suspension bridge in the United States, which links Brooklyn to Staten Island, where I am headed. There, just about a dozen miles from the tip of Manhattan, looms the largest man-made structure in the world: the Fresh Kills landfill.

There are two man-made things it's said you can see with the naked eye from outer space: the Great Wall of China and Fresh Kills.

It's obviously a massive dumpsite—more than 2,200 acres, or about five square miles. To put that in perspective, it is about three times the size of Manhattan's Central Park. It holds more waste than anywhere else on land in the world.

From 1948 until 2001, Fresh Kills took in refuse from America's largest city—and piled it high—so high that it is the tallest point on the Eastern Seaboard of the country. Not many people know about Fresh Kills, which is surprising. No one tries to hide it. In fact, you can drive straight through it on New York State Route 440. There, you'll see a parking lot of trash trucks, mounds of waste, awesome machinery and equipment, and even a little green "Fresh Kills" sign letting you know you are there. But you won't see anything being hauled in.

Fresh Kills ceased operation because the volume of waste was getting out of hand. After years of protests—not surprising considering that a half million residents live near it on Staten Island—the city was finally convinced to shut it down in March 2001.

Waste is now dispersed to different landfills throughout New York, and some is even shipped out of state.

Fresh Kills reopened temporarily to receive World Trade Center wreckage after 9/11.

Some 2 million tons of wreckage—and, yes, tragically, human remains and personal items—were shipped to the site, adding to the tens of millions of tons of refuse that had already been rotting there for decades.

The gates were then shut for good.

Now, there are plans to make Fresh Kills one of the world's biggest parks. On certain days, public tours are given.

I skipped the tour option.

THE PERIMETER OF Fresh Kills is fenced off. I drive around its borders. Barbed wire in many places sits atop the fencing. There are signs, "Danger Hazardous Waste," "No Trespassing," "Warning."

I park on a side street and walk across the parkway to gaze through the wire barrier at what lies behind it. I notice just a few feet away that someone has cut a small opening in the fence—large enough for me to slip through. There's a thicket on the other side of the fence blocking my view beyond. So I risk slipping to the other side.

I make it into the thicket, but beyond it is a small mound from which I think I'll have an even better view. I take the few steps to the top, but misjudge the steepness on the other side, trip over a vine, and fall ass-over-teakettle down the other side. It's a grand tumble, the kind you wish you had on videotape: the expression, "oh, yes indeed, I am going down," dawning on my face, the "better put my hands out in front of me to protect my landing" coming too late so it's just a face plant into the swamp. Thud. I'm on the ground. I am one of the few people in the world now, I believe, who can honestly say he's eaten dirt at the world's biggest garbage dump.

Anyway, the ground here isn't like the ground at other places, like, say, a park. At a park, there is a layer of grass, under which there is topsoil, then subsoil, and under that, substratum material before heading down further into groundwater and deeper into the Earth's crust.

Landfills like Fresh Kills layer their waste. At the top, there is usually a layer of soil, and then a covering cap made of clay or plastic, then the garbage or trash "cell," then another layer of sand or gravel, then older cells of trash, and then a bottom liner before the groundwater. In between there are, depending on the sophistication of the landfill, methane gas recovery and storm water drainage systems.

My face is planted in the top layer of the site. I'm guessing this is where the plastic bottles were buried because there are still hundreds of them strewn about.

Landfills separate trash into groups: plastics, papers, metals, foods, yard trimmings, etc. The groups are where the recycling process begins and ends. But we'll get into that later.

Right now I want to talk trash. As the Environmental Protection Agency defines it, our trash is made up of the things we commonly use and then throw away. These materials range from packaging, food scraps, and grass clippings, to old sofas, computers, tires, and refrigerators. Trash does not include industrial, hazardous, or construction waste.

Waste from our homes amounts to between 50 and 65 percent of the total waste generated. Waste from schools and commercial locations, such as hospitals and businesses, amount to between 35 and 45 percent.

Most municipal, meaning government-run, waste sites such as Fresh Kills are 34 percent paper; 13 percent yard trimmings; 12.4 percent food scraps; 11.7 percent plastic; 7.6 percent metals; 7.3 percent rubber, leather and textiles; 5.5 percent wood; 5.3 percent glass; and a few percent of other materials, according to the EPA.

For every ton of paper we toss, about half of it gets recycled. For every ton of plastic dumped, about one-tenth gets recycled. And for every ton of glass, about one-quarter gets recycled. That annual recycling rate equates to about one ream of paper per week for every female in the country; a two-liter plastic bottle per day in the hands of every baby boomer; and a glass jar every day of the school year for every kid under the age of fifteen.

The EPA and others point to what a great job we're doing with our trash. We are. But there is something more to the picture.

We do a good job of recycling—better than in the past—but we could do even better. Since 1960 the amount of waste we each produce has almost doubled, from 2.7 pounds to 4.6 pounds per day. During that time the US population has almost doubled as well: from 183 million people in 1960 to more than 300 million today. More people, more waste. Meanwhile, our recycling rate has skyrocketed. If you look at the raw numbers, we recycle about fourteen times more than we did fifty years ago. However, over the same period of time, 78 percent of the nearly 8,000 landfills that were open in 1960 have been shut down, because they either reached capacity or failed to meet national health and safety standards. At first glance, the relatively small number of landfills in operation today—a mere 1,754—may appear to indicate that we're dumping less waste overall. We're not. Something that isn't so well-reported is that landfills have gotten bigger over time—about twenty-five times bigger. Just a decade ago the average landfill's capacity was 1 million tons of trash. Today, it's 25 million tons. So while it may seem like we've made progress in our waste generation by reducing the number of landfills, they have significantly increased in size. There may be fewer of them, but they are much more massive.

The latest data available from the EPA show that Americans generate about 251 million tons of waste per year, 82 million tons of which are recycled. If you're like me, you do some quick math and say, "Great, but that still leaves 169 million tons of garbage."

Except for 1990—when the EPA began a national campaign to achieve a 25 percent goal of waste-reduction and recycling—we have been sending more total waste to landfills now than ever. And that's where we can do better.

Do we, for example, need to use so many disposable plastic water bottles? We toss out more than 81 million of them every day. Yes, every day. If those of us who drink bottled water could cut our use in half by refilling each bottle once before we tossed it, we would save more than 1 billion pounds of plastic from being sent to landfills each year.

Or what about plastic bags? Is it necessary to use two to carry a single item? About 100 billion plastic bags are tossed per year in the United States alone, never mind the rest of the world. If you reused just two bags per week, that would add up to over a hundred per year. Extend that to every US citizen and the number of plastic bags discarded would be cut by one-third.

We can and should accentuate the positive news about recycling efforts. Already, by recycling nearly 7 million tons of metals (which includes aluminum, steel, and mixed metals) a year on average, we eliminate greenhouse gas emissions totaling close to 6.5 million metric tons. The EPA says,

> Recycling has environmental benefits at every stage in the life cycle of a consumer product—from the raw material that it's made with to its final method of disposal. Aside from cutting greenhouse gas emissions, which contribute to global warming, recycling also reduces air and water pollution associated with making new products from raw materials. By seeing used, unwanted, or obsolete materials as industrial feedstock or new materials or products, we can each do our part to make recycling work.
>
> Nationally, we recycled 82 million tons of municipal solid waste. This provides an annual benefit of 49.7 million metric tons of carbon equivalent emissions reduced, comparable to removing 39.4 million passenger cars from the road each year. But the ultimate benefits from recycling are cleaner land, air, and water, and overall better health.

The problem of waste and what to do with it doesn't exist just in the United States, of course; but in the world over. Brazil generates so much plastic bottle waste that they actually make park benches out of them. Recycling helps reduce waste in India, as we've seen. And in China, composting is an integral part of the agricultural and farming system.

In terms of landfill space, we could save 75 percent of the land space we currently use for trash if we only recycled more. In the United States, that land space is the size of Pennsylvania.

To be sure, recycling has caught on. We've got the hang of putting our paper in one bin, our plastics in another, and our yard trimmings in yet another.

If we didn't recycle at all, this country would have another 2 billion tons of total trash to deal with. If it were compressed, compacted, and all piled together in one place—let's say California—it would cover all 4,000 square miles of LA County to a depth that would reach the waists of the top NBA players. Clearly, this is not a viable disposal option. So, to hold this extra waste, we'd need at least eighty more of those gigantic 25-million-ton-capacity landfills, and we'd have to build another three or four landfills each year to hold the additional 80 or so million tons of materials that we'd be throwing away instead of recycling. And because, for some reason, people don't want landfills built in their backyards, new landfills would need to be located in increasingly remote areas. This would only make for higher disposal costs, greater diesel fuel consumption, and recurring trash fee hikes for residents as more and more garbage is transported to middle-of-nowhere dumpsites.

This is only part of the picture, because recycling isn't just about reducing waste. It's also about reducing the use of virgin materials.

If we didn't recycle paper, for example, we'd need an additional 340 million trees and some 40 million barrels of oil per year to make an extra 20 million tons of paper from wood pulp. If we didn't recycle aluminum cans, we'd need another 35 billion kilowatt-hours of energy per year—equivalent to building ten new coal-fired power plants—because making cans from recycled aluminum uses only a fraction of the energy needed to produce aluminum from bauxite ore. Not to mention that the mining of that bauxite would generate an additional 4 million tons of toxic waste annually—which, of course, would need to be put somewhere.

In a practical sense, when we don't recycle, waste piles up like Fresh Kills, four mountains made from 150 million tons of solid waste just sitting and degrading. This is our reminder.

There are, of course, alternatives to waste disposal besides recycling and landfilling. We incinerate about 12.5 percent of our waste, most of it paper from discarded packaging. And we ship our waste elsewhere, sometimes without much success.

There is the famous story of the ship named *Khian Sea,* which in 1986 was relegated to circle the globe for fourteen months, filled with waste from Philadelphia and being refused entry to eleven countries on four continents. Nobody wanted the ship's cargo and nobody knew what to do with it. Then, somewhere between Singapore and Sri Lanka, the ship's frustrated crew had an idea; and the waste, much of it hazardous, mysteriously disappeared. The ship was then allowed port. The crew had no comment except to say that one day the waste was there, and the next day it was all gone. Hmm . . .

Figuring out what to do with our waste is a huge undertaking, and it's big business. The waste management industry is a $40 billion market. Some estimates put the recycling industry at nearly $250 billion because of the new products that are made. There's so much money in waste that people and businesses get creative. There have been ideas to jettison our waste into outer space, or deep into the oceans, or even to dump canisters into the Antarctic snow and letting it melt its way deep into the ice cap. Genius idea, that one.

Still, the most common way to rid ourselves of our refuse is just to dump it on land. That's how Fresh Kills got so big.

In 1948 Fresh Kills was opened to deal with New York City's mounting trash problem. Until Fresh Kills, trash had been shipped to more than a dozen landfills via train, barge and truck as far away as Pennsylvania. Fresh Kills was supposed to be only a temporary solution, so it was not built with longevity in mind. It was not until decades later that liners were installed to prevent leakage into the groundwater. Air filters were lacking too, and nearby residents complained of the stench. Lawsuits were filed, and in 1991 plans to reduce reliance on Fresh Kills were put in place, mostly because the EPA enacted laws at the time limiting landfill capacity and monitoring surrounding air and water quality. But by the 1990s much of the damage had already been done. At the height of its operations in the 1980s, Fresh Kills was taking in 30,000 tons of waste per day. The average landfill in California, by contrast, accepts 1,500 tons a day.

Tests show that waste, if left untreated, emits all sorts of hazardous pollution into the air and water.

This is how it happens: when paper, yard waste, kitchen scraps, and other materials made from once-living things are buried deep in a landfill, they don't decompose easily due to lack of oxygen. Yet that doesn't mean buried trash never decomposes. Certain types of bacteria thrive in anaerobic (oxygen-free) environments. When these anaerobic microbes munch on organic garbage, they exhale a biogas mixture of carbon dioxide and methane.

In ideal conditions for decomposition, the paper you toss takes between 2 to 3 months to biodegrade; a banana peel out the window a few weeks; a pair of old socks up to 5 years; leather shoes 50 years; a plastic six-pack holder 450 years; and an aluminum can about 200 years. All the while, the degradation process is taking place, gas is being emitted—gas from your stuff. Your remnants end up as solid waste in landfills, but they also take flight. Landfills are the largest man-made emitters of methane in the country.

This is another reason to recycle: when we don't give an item the opportunity to biodegrade, we reduce the amount of pollution that's emitted.

Note that there is a difference between recycled-content goods and post-consumer recycled goods. Recycled-content products are usually made from the leftovers of production when new goods are manufactured. With paper, this is the extra shavings of pulp and sawdust, for example. Post-consumer recycled products, on the other hand, are made from materials that have served their function for consumers and have been recovered through a recycling program. The latter is obviously better for the environment because virgin materials are being saved.

In any case, when we use something and toss it, it ultimately ends up coming back to us in the form of a new product or in the air we breathe as it decomposes. We never really rid ourselves of what we use. It doesn't just disappear. What we toss goes somewhere, and over time, that somewhere eventually becomes here.

Because of its mass, Fresh Kills was once reported to emit more than 5 percent of all methane gas emissions in the United States, and about 2 percent of the entire world's. Air studies at Fresh Kills show pollution traveling to Brooklyn, Manhattan, New Jersey, and beyond—much of this occurring merely from trash decomposing.

Many landfills also incinerate items. When they do this, gases from incinerated garbage waft high into the air and travel with wind currents as smoke and ash. If an item is recycled, however, it ends up compacted, remixed, and reused.

If you've never been to a recycling center, picture this: large cylindrical containers are lined up out front so people with small quantities of things can just drop them off: aluminum cans and foil, tin cans, clear glass, green glass, brown glass, junk mail, plastic bottles, newspaper, cardboard.

TOP: A single stone lies at the physical vortex of the three major faiths on Earth, and that rock too is crumbling.

LEFT: Half the planet's population is connected to Jerusalem through their religious beliefs.

BOTTOM: The world's monuments are decaying faster due to climate change.

TOP LEFT: Without recycling, Mumbai would be overstrewn with waste.

LEFT: Clean up attempts seem futile, but are increasingly urgent as population rises—a portent for the world.

BELOW: We often use developing countries as waste grounds for the West; 85 percent of US electronic and hazardous waste is shipped overseas.

TOP: On Borneo Island, illegal logging is rampant.

CENTER: Timber operators clear land to plant palm oil, which we in turn use in our toothpastes, soaps, and as a common food ingredient.

BOTTOM: Meanwhile, the most species rich place on Earth is devastated, corrupting a virtual Eden.

RIGHT: In Linfen City, China the smog is sometimes so thick it's difficult to see more than a few yards in front of you.

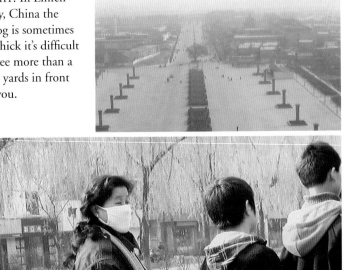

ABOVE: Just breathing the air everyday is as harmful as smoking a pack of cigarettes.

LEFT: A coal-fired plant gets built every four days in China, stacking the pollution problem worldwide.

LEFT: Seal hunting is crucial to the 4,000-year-old subsistence culture in Shishmaref Village, Alaska.

ABOVE: But the ice isn't forming its usual hunting grounds; waves crash to shore.

RIGHT: Coastal erosion wrecks homes, pushing them into the sea and is forcing the entire village to move.

ABOVE: The Amazon proves the law of the jungle true. There is a whopping 45-feet difference in water level between wet and dry seasons.

CENTER: The Fresh Kills landfill is the largest in the world.

BOTTOM: It's being made into a park with views of New York City.

LEFT: A floating waste site twice the size of Texas is culled by nature.

CENTER: It captures man-made refuse in ocean currents.

BOTTOM: The world's litter washes up on shore from afar in Kamilo Beach, past the southern most point of the United States.

LEFT: The Great Lakes help form the largest surface freshwater system in the world.

CENTER: Ships from as far away as Asia dock 2,300 miles inland from the Atlantic.

BOTTOM: At home much of our waste and its effects circle right back to us.

The recycling process itself is rather simple. Adam Holt, manager of the Allan Company recycling center in Santa Monica, California, which is my own local facility, explains how it works: "We acquire it, process it, and market it." The "it" he is speaking of is trash. "You come in, put what you have on a scale, weigh it, take your receipt to the register and get paid," he says. Yes, paid. Many people are surprised to learn that beyond the deposits they can get back on bottles and cans, they can also get paid for all sorts of other stuff: cardboard, metals, paper, even stone debris.

"We want to get the word out: bring us your stuff and get paid. We need more trash," Adam says.

He gets truly excited when he talks about recycling: "I watched this segment on ESPN last night where a guy in Minnesota started collecting cans to help get more money for the high school athletic department. They didn't have enough money in the budget for uniforms. When the word got out, everybody in town started collecting cans and recycling and donating the money. It was awesome!"

Allan Company pays anywhere from thirty-three cents to almost four dollars per pound for different types of material. When there is enough for a "clean" truckload—45,000 pounds of material—it gets transported to a mill—where it is processed and made into fodder for other products of its kind—or it is sent via train to port where it ends up at a mill in another state or another country.

I notice several workers at the facility sweeping up the excess material left on the ground after a truckload leaves and then re-sorting it; nothing seems to go to waste.

It isn't always easy to figure out what you can and should recycle and what you shouldn't. Municipalities have different rules for which items are recyclable and which aren't. That means you have to contact your city or local waste hauler to figure out what to do.

Still, most recycle bins come with a label listing acceptable items. On my bin, for example, it says, "Recycle Here" and then provides a list: cans, glass, plastic, mixed paper. And then it gives specifics: glass beverage containers (all colors); plastic beverage containers; all colored plastics numbered 1 to 5; all steel cans; all aluminum beverage containers and cans; catalogs, magazines, office paper, junk mail; brown paper bags, telephone books, newspaper; cereal boxes, corrugated cardboard (flattened). Then it says, "Do Not Recycle Here." No lightbulbs, ceramics, plate glass, Pyrex, plastics #6 and 7; no Styrofoam, hazardous materials, batteries, aerosol cans; no coffee mugs, wire hangers, household appliances. Then it lists a telephone number to call for the city's recycling division.

I want to know what I can do with my wire hangers besides return them to the dry cleaner, and my batteries. I call the number posted on my bin.

The battery disposal instructions are easy: take them to the materials processing center about three miles from my house and the city will take them off my hands and dispose of them properly, I'm told. "How?" is a question outside the realm of knowledge of the person with whom I'm speaking. I gather that they're likely shipped off somewhere overseas. There isn't much you can do with the heavy metals, such as mercury and lead, which batteries contain. Some outfits do extract metals and batch them for other uses, but the environmental solution is really just to keep those metals from coming into contact with air, water, and soil. The wire hanger question poses some confusion, for the city recycling official as well as for me.

"Oh, you can just put them in the recycle bin."

"But it specifically says not to," I say.

"No, it doesn't."

"Yes . . . it does; I'm looking right at the label on the blue bin."

"Then the label is wrong. We accept wire hangers for recycling."

I'm not alone, I know, in getting such circumspect answers. However, the point is that thousands of people are being misled in my city alone by erroneous recycling information.

The plastics numbers also give me pause. Sometimes the plastics recycling code isn't so apparent, so you have to look for it. The plastics recycling code is that number in the center of the triangle with arrows for sides (the recycling sign).

Each number corresponds to the type of resin used in making the plastic. The plastics industry came up with this labeling system to help people identify which plastics are used in different types of products. Number 1 is commonly found on soft-drink and water bottles; number 2 on detergent bottles and milk jugs; number 3 on plastic PVC pipes and outdoor furniture; number 4 on dry-cleaning bags and trash-can liners; number 5 on bottle caps and yogurt containers; number 6 on Styrofoam peanuts and cups; and number 7 on every other type of plastic.

These numbers make certain products easier to identify for recycling purposes, and help people determine whether the container they're holding is accepted in the recycling bin or should be tossed into the trash.

Most municipal recycling programs in the United States accept only #1- and #2-type plastics. And not all of those acceptable types of plastics are actually recycled.

This is a common misconception about recycling: just because an item *can* be recycled doesn't mean it *will* be recycled.

Don Avila, the public information officer at the Puente Hills landfill in Whittier, California, which is now America's largest operating landfill since Fresh Kills closed, says when things are just dumped into a pile and then separated after the fact, it's likely that materials that could be recycled are not.

There are dirty MRFs (materials recovery facilities) and clean MRFs. Dirty MRFs are where trash is piled together and dumped and then separated, he explains. This is how material that could be recycled gets lost in the shuffle. Clean MRFs are where recyclables are separated before by the consumer. Clean MRFs rely on people to do the right thing with their recyclables and put them into the proper bins at home. "People need to be more conscious," Avila says. "If they are, we can reduce the size of our landfills, lower the costs of disposal, and eliminate health hazards."

An average disposal fee for one ton of trash is about forty dollars. Depending on where you live, this fee is built into local taxes or you're billed directly by your waste hauler.

Confusion about what to recycle and how to do it still abounds. Just 7 percent of all plastics are recycled, for example. Maybe that's why there are still so many plastic bottles at Fresh Kills strewn about.

THE WEEDS MAKE it difficult for me to see anything. My first step is on another bed of plastic bottles. They're old, in a crackly state of decay. I walk over them to what appears to be an opening about ten yards away. The bed of bottles extends. Crunch, crunch, crunch. I walk until I find myself on top of a concrete patch. In front of me is a small pond. Seagulls circle overhead. There is no one about. It's spooky even though it's broad daylight. This looks like the perfect place to bury a body. Curious why that thought enters my mind—maybe because the fictional home of the Sopranos is just over the bridge in New Jersey.

I look that way, toward Jersey. In the distance, black and white smokestacks loom. Nothing comes out of their tops. They aren't in use. I sniff and get a whiff of that sweet stink that exists at garbage dumps. I can hear machinery in the distance even though I've been told the landfill is closed. I walk the grounds circling various estuaries that can be found at Fresh Kills and get up close to the machinery on the hill. Trucks are moving about large piles of junk. Fresh Kills isn't accepting new garbage, but it's still dealing with the old stuff.

A modern landfill operation probably wouldn't exist so close to a city as Fresh Kills does. The environmental effects are too profound. Most waste today gets picked up at your curb, and recyclables are brought to a materials recovery facility where they are sorted, organized into different groups, packaged, and delivered to industries that can make new products from them. In many places, nonrecyclable trash is loaded into big metal containers and trucked to what is called an intermodal facility, or transfer station. This facility loads the sealed containers of trash onto train cars so they can be carted to a remote landfill. Each container can hold as much as thirty tons. Each train can haul 130 to 200 containers depending on the amount of trash in each container. That's enough for a nice-sized city's worth of daily waste. The containers are removed from the train at another intermodal facility and emptied into the remote landfill.

The difference between a dump and a landfill is that a dump is exactly that: an unregulated hole in the ground where garbage and trash are tossed. A landfill is designed to meet environmental standards. Those standards include liners and gas-capture systems, which are intended to prevent pollution from leaking into the groundwater or entering the air.

Maintaining a landfill is complicated stuff. Dirt has to be watered down in dry seasons so dust doesn't kick up and carry toxins. Drainage systems have to be installed so water runoff doesn't seep into surrounding areas and the underground water. Groundwater moves slowly underground, generally at a downward angle (because of gravity), and may eventually seep into streams, lakes, and oceans.

Because water percolates through the cells and soil in the land-fill, much like it percolates through ground coffee, it picks up contaminants (organic and inorganic chemicals, metals, and bio-logical waste products of decomposition) just as water picks up coffee in the coffeemaker. This water with the dissolved contami-nants is called leachate and is typically acidic. To collect leachate, perforated pipes run throughout the landfill. These pipes then drain into a collection pond where the leachate is chemically treated like any other sewage or wastewater.

Methane is also captured at many landfills through a set of pipes embedded within the layers. The primary reason for this is to prevent dangerous gas buildups and reduce air pollution. But methane—known also as natural gas—can also be sold for use as fuel.

As mentioned, none of these techniques was contemplated when Fresh Kills was first operating. Later attempts were made to install more modern sanitary features, but neighbors and city of-ficials claimed it was too late. They believed residents had already been exposed to toxic fallout from Fresh Kills.

In 1995, an extensive emissions study confirmed that the Fresh Kills landfill released more than a hundred organic chemicals into the air. Driving, sorting, and general operations were also found to kick up dusts that contained metals and other toxic chemicals. Some of this certainly made its way into the lungs of area residents. And because of the long decaying process for particular items, Fresh Kills likely will release organic chemicals, metals, and other pollutants into the air for years to come.

Remember Rod, the guy from Queens who went over to Linfen, China, to teach English? Well, Rod used to live next to Fresh Kills. He told me that, after I interviewed him and in-formed him that I would be headed to the site.

"Man, that place. I used to walk out of my backyard and through the woods and onto the grounds at night. You could see all these piles so high," he recalls, his eyes growing wide. "Then after 9/11, a convoy of trucks came through town with all this stuff blowin' off the back. That couldn't have been good. I'm surprised it didn't glow."

Shortly after that, he moved. He said there was always talk in the neighborhood of the toxic fallout from Fresh Kills and people complaining of the smell and getting sick. "Waddya gonna do?" he shrugged.

What we can do is reduce, reuse, and recycle. They are the three *R*'s so often referred to by environmentalists. What they call for is to reduce the amount of products we consume and throw out; reuse what products we can; and recycle whichever products are allowed.

If you just recycled your Sunday papers, for example, you'd save four trees a year. The average person also has the opportunity to recycle 25,000 aluminum cans in their lifetime. Recycling a single aluminum can saves enough energy to power a TV for three hours. It's something to think about next time you're watching the game; one can could power your view time.

The word *kills* in Fresh Kills is derived from the Dutch word *kille,* which means riverbed or water channel—not the other definition. Fresh Kills was largely built on marshlands. It is a watershed basin that drains much of the wet lowlands of Staten Island into the Arthur Kill, or channel. Streams and branches run from the most western portion of the island to its southern side. The land, all around, is famously wet.

There is nothing particularly special about the ground on which I'm walking. It's what lurks beneath that counts. In fact, it's quiet here, almost peaceful. The wind rustles the weeds. Birds glide. When they flap their wings you can hear it. Otherwise it's still.

As I get back into my car, a minivan I borrowed from some friends in New York City, one of the residents on the street comes out of his house. He hacks and coughs before getting into his truck, which has an advertisement for landscaping on the side panel. He could be sick from any number of causes, but to me it's a reminder of the hazards landfills pose on health.

Many health hazards come from living near a landfill. One study shows that concentrated landfill gases may travel up to a mile or more from a landfill site, particularly downwind. The most dangerous and pointed-to toxins are the volatile organic compounds (VOCs), which have numerous chronic health implications: from cardiac, reproductive, and birth defects, to developmental disabilities. More immediate symptoms include nausea, vomiting, headaches, shallow breathing, and coughing, as well as irritation in the eyes, nose, and throat.

The city of New York, however, wants to transform the legacy of Fresh Kills. The Fresh Kills Park Project is focused on converting the landfill into a new public park over the next thirty years. It dubs the project "one of the most ambitious public works projects in the world, combining state-of-the-art ecological restoration techniques with extraordinary settings for recreation, public art, and facilities for many sports and programs that are unusual in the city."

The parks department exclaims that "the tops of the landfill mounds themselves offer spectacular vistas of the expansive site, as well as views of downtown Manhattan."

Think about that for a second: standing on top of a mountain of trash gazing at the New York City skyline. Are these really the vantage points from which we'd like the next generation to look at the world? Is this the mounting testament to our place on the planet—consumption and debris?

A new park may come out of Fresh Kills. This, don't get me wrong, is a good thing. I celebrate the fact that there is such a mad cleanup effort around such a monstrous refuse site. Awareness may yet come when people are exposed to the enormity of Fresh Kills and can walk its grounds.

I drive to another part of Fresh Kills and stand tall on a concrete slab and gape at the miles of land ahead of me. Fifty percent of the Fresh Kills site is garbage. It is underneath the ground beneath my feet. It is there in front of me. It is there in the distance. I look down and see a small, black plastic comb. I try to imagine all the fragments of lives that Fresh Kills stores—millions and millions of peoples' expelled belongings are here. Socks, hats, clothes, love letters, pieces of furniture, jewelry, mementos of existence fill the land. It's a crying shame how much we consume and throw away with abandon.

I have been careful throughout this chapter to state that Fresh Kills holds more waste than anywhere else *on land* in the world. The sea is a different story. There, way out in the middle of the Pacific Ocean is a garbage patch that makes Fresh Kills look like a small blip. This is where we go next: the Eastern Garbage Patch, the largest refuse site on the planet—a floating mass of trash.

EIGHT

Where the Currents Take Our Trash

The Eastern Garbage Patch, Pacific Ocean

In the middle of the Pacific Ocean drifts a garbage patch twice the size of Texas. The circular rotation around it draws in trash like a vortex. Flotsam and other debris combine to form huge floating clouds of waste. This waste comes from cargo ships: eighty thousand pairs of Nike sneakers, tens of·thousands of rubber duckies—yes, rubber duckies, bobbing around since a cargo spill in 1992—the odd, or should I say odder, disgorge of hockey equipment from yet another spill. All this mashed together with plastic bottles, tops, six-pack holders, and other litter that degrades into smaller and smaller fragments as it is exposed to the elements; bite-sized pieces for birds and fish that eventually die from ingesting them.

The Eastern Garbage Patch is a lethal marine habitat that has grown and expanded over decades.

It isn't just cargo ship mishaps that cause these vast waves of waste in the North Pacific Gyre. We contribute to it too. Around 60 billion tons of plastic are produced each year, about 10 percent of which ends up in the sea. About 20 percent of this is from ships and platforms, the rest from land. In other words, about 80 percent of the trash that ends up in the ocean comes from on-shore. The wind carries it, sewage pipes spill it, even our garbage disposals make ways for waste to enter storm water drains and to eventually flow out into the ocean.

Take a walk along the beach anywhere in the world and you'll find plastic bags, bottles, and containers. Along with traffic cones, disposable lighters, old tires, and toothbrushes, these items have been casually tossed away. From the shore, they get carried by wind and tide to the sea. Currents bring them here, to the largest dumpsite in the world, where they join the mass of plastic, paper, oil, rubber, wood, rope, fishnets, and virtually every other type of material on the planet.

We've already seen how pollutants from local landfills contaminate air, soil, and groundwater. But the oceans are vast. And when toxins pollute the sea, there is potential for even greater environmental hazard. Oceans occupy 70 percent of the Earth's surface and are home to over 90 percent of all life on the planet. Seafood is the primary source of protein for many coastal peoples. Worldwide, nearly a billion people rely on fish for a big source of their daily food protein. When we pollute the water, we pollute the fish.

The Eastern Garbage Patch through which I am sailing off the coast of Hawaii is contributing to the ocean's demise. Rivers, streams, and sewage pipes propel waste out to sea, and it eventually ends up here. This is the place about which we sometimes wonder: "Where does all the sewage go that's pumped out into the ocean?" In this area there are about a million pieces of garbage within every square mile, according to some estimates. Currents pull and drag garbage in this direction, not far from where pirates, I'm told, used to search for bodies fallen overboard. They understood the movement of the currents, and knew where to find drifting loot and corpses. This is also near where the first Hawaiians are said to have landed from Polynesia.

Hawaii acts as a comb for the Garbage Patch. Nineteen islands and atolls comprise the Hawaiian Islands. They sit smack between North America and Asia in the middle of the Pacific Ocean and are the most remote islands on Earth. The Eastern Garbage Patch is estimated to float between the coast of California and the Hawaiian Islands; and there's a "superhighway" from there to another ocean garbage patch just south of Japan, the Western Garbage Patch. That patch collects trash from Asia, Russia, India, and the Malaysian Peninsula and deposits tons of trash on the south coast of Japan; its concentration of debris is said to be even higher than what I am seeing here.

The superhighway acts like a funnel connecting the two garbage patches. It's how whiskey bottles from Japan, pill bottles from India, Korean detergent containers, and oil cartons from Guatemala make their way to the Hawaiian shores. Those are things I saw firsthand.

Captain Charles Moore has been investigating the Eastern Garbage Patch since 1997, when he stumbled across it while speeding across the ocean during the Transpacific Yacht Race. The TransPac race starts in Los Angeles and ends in Honolulu.

"Usually, you go in a straight line and you don't go through the Garbage Patch. But there was an El Niño in '97, and a lot of debris had floated south, so I noticed it." He relates this story to me as we sit on the deck of his boat, the *Alguita,* early one morning in Hilo, Hawaii, as the sun rises in front of us.

He says the gyres—giant, circular oceanic surface currents—calm the waters between Hawaii and California; so much so that yachtsmen typically sail north toward Oregon or Alaska to catch enough wind before tacking south. The calm waters are known in the sailing world as the "horse latitudes," because it was here in times past that ships would jettison their horses overboard to lighten their loads and make better headway in the calms.

At any rate, Moore had some extra fuel that year and decided to take some time on his way back to California to trace the scores of bottles, caps, and plastic bags he was seeing in the ocean. "It's a big blue ocean so you expect to see some trash part of the time. But for a whole week we were seeing it," he says. The Garbage Patch is spread out and debris is dispersed on the ocean's surface as well as underneath.

"It isn't what people think. It isn't some pile of garbage that you can land on and see all of the time," Moore says. But you can see garbage much of the time, for miles and miles. That piqued Moore's interest. So he started doing some investigating, taking samples he found along the way.

Moore is the go-to expert when anyone investigates the Eastern Garbage Patch because he has been the most active in trawling it.

Before he stumbled across the patch, Moore had been conducting water-quality samples along the West Coast of the United States, examining areas of pollution, mostly from where rivers meet the sea. Little did he know how far the pollution extended.

The Eastern Garbage Patch begins to take hold about a thousand miles off the California coast. It's extensive and it's spread out. It's filled with myriad materials, but mostly plastics. And as we all know, plastic does not biodegrade but rather breaks down into tinier and tinier pieces; it doesn't go away. For example, there are enough particles of plastic in just one liter-sized bottle to put a piece on every mile of beach in the entire world. Those are the tiny particles that are ingested by fish and other sea life. Now think about those 129 million plastic water and soda bottles we discard every day. Those alone are enough to make an ocean of plastic, or what Moore has come to call "the synthetic sea."

Moore has founded a research institute, the Algalita Marine Research Foundation, to study plastics' effects on marine life. His passion stems from growing up near the ocean in Long Beach, California, and sailing with his father, an industrial chemist, to remote locations such as Guadalupe Island. He and his family are heirs to the Hancock Oil company.

Moore majored in chemistry in college, but ran a woodworking and finishing business for twenty-five years before launching the *Alguita* in Hobart, Tasmania. He had the *Alguita,* a fifty-foot catamaran, specifically built for conducting research. Moore also launched the Australian government's first "Coastcare" research voyage to document the contamination of Australia's east coast.

With salt-and-pepper hair—a dash more salt than pepper— Moore looks the part of sea captain. When I met him for the first time at a restaurant, I didn't have to guess twice about who he was when he appeared at the entrance in a crowd of people. The sea's effect on the lines of old sailors' faces portrays the distances they've traveled and the gravity of their respect for the will of nature over their own. Besides, he walked, as a good shipman does, like a monkey.

Moore's findings, summarized in a 1999 study, shocked the scientific world. He found six times more plastic fragments by weight in the central Pacific than associated zooplankton. Here's why that study is worrisome.

Zooplankton are a type of plankton—organisms that live their lives as underwater drifters. *Plankton* means "drifter" in Greek, and *zoo* comes from *zoon,* which means "animal." Plankton can be all different types of organisms, from single-celled algae to krill, to jellyfish. Plankton make up the base of the marine food chain. That is, they influence the survival of everything from fish to crabs to sharks to sea lions. They're some of the most important organisms on Earth. Zooplankton are nonselective feeders, taking in anything small enough to ingest. Ideally, they'd be eating nutritious phytoplankton and algae. But given that there's six times more plastic floating around than zooplankton, it's inevitable that a significant proportion of the diets of zooplankton will be plastic. Now, when you consider that a marine mammal like the blue whale consumes between 4,000 and 16,000 pounds of zooplankton per *day* of the summer feeding season, we've got a big problem on our hands.

In the scholarly journal *Marine Pollution Bulletin,* nine scientists supported many of Moore's findings with their own aerial and satellite study of the patch. Disturbingly, they observed animals "preferentially foraging in the same convergence zones that concentrate marine debris, thus increasing their risk." During the winter and spring, when the patch is most concentrated, fishing boats from Hawaii head out to it because that's where the fish gather.

Let me now spell out the point: when fish eat plastic and we eat fish, we end up ingesting plastic too; the trash we toss ends up in the fish we eat. It comes back to us this way. As a reminder, plastics can hold toxins that are extremely hazardous to our health. And most of those toxins can't be cooked out.

According to the US Department of Health and Human Services, certain chemicals found in plastics when ingested can cause nausea, abdominal pain, loss of appetite, joint pain, fatigue, and weakness, as well as skin disorders, nervous and immune system effects, as well as effects on the liver, kidneys, and thyroid gland. A battery of other ailments and disorders can ensue depending on the chemicals associated with the types of plastics a body ingests. And there are a lot of different types of plastics floating around at sea.

Some of the odder things Moore has run across in the middle of the ocean include cathode ray tubes for televisions, basketballs, toothbrushes, and hockey sticks (that tanker spill is hard to forget). Mostly, Moore finds lots of fishing nets, many of which are also made of plastic. Fishing boats can haul nets twenty miles long in some places and oftentimes nets, or parts of nets, break away.

Moore trawls with two plankton nets that are designed to catch minuscule pieces of material. They are attached to cylinders several feet in diameter. When the cylinders are tossed overboard, the nets stretch out like balloons and capture material—plastics, fish, seaweed, or whatever else is floating in that patch of ocean. The material is brought back on board, tagged, and stored.

As we sail the *Alguita* off the coast of Hawaii, several hump-back whales cruise past. They spout. One of the crew wonders about the effect of plastics on them too, and whether the whales will be long for this area. There is so much trash that floats to shore on Hawaii that Moore's Algalita Institute considers Kamilo Beach, on the southern tip of the Big Island, the most polluted beach in the country. Kamilo is near South Point, the southern-most point of the United States.

To get to Kamilo Beach you have to do something that isn't environmentally friendly—take a four-wheel drive. The beach is located miles off the main road, just past a series of wind farms.

Jeff, a recent graduate from the University of Hawaii at Hilo, crews for the *Alguita,* and knows the secret road you have to take to get to Kamilo. He is willing to serve as our guide. I am joined by Dr. Marcus Eriksen, Algalita's director of research and educa-tion, as well as two additional Algalita staff members to sample findings.

After hours of driving, we stop in front of a chained gate. Beyond it lie fields. Ranches are off to the left and right. Cattle and horses roam. There is no visible sign of a road past the gate, but Jeff, an energetic surfer dude, hops out and unwinds the links; the gate isn't locked. The chain is there to ward off those who aren't in the know, Jeff says. Surfers come here to catch big waves.

Bump. Crash. This is all rock, rubble, and dirt. The "road" looks like a mogul run. And this, Jeff says, is the easy part. "Wait until we get to the lava rock."

The rented—especially for this trip—Chevy Trailblazer is mine and I flash back to a question that gnaws at me the entire bumpy, scratchy, stuck-in-the-sand way: "Would you like to add insurance for $5.99 a day?" My answer "no" echoes in my head.

We go through three more tied-up but unlocked gates and bottom out several times before we hit the lava patch. It's ugly. It's the sound of steel scraping rock the entire quarter mile or so, which takes a good half hour to traverse. All passengers are out to lighten the weight.

Past the rocks, sand is a welcome relief. It's remote and so, so beautiful here. Mountains rise in the distance. The sun is out. The blue sky is wide. We cruise along the dunes and hang a sharp right toward the ocean on Jeff's direction. We're stopped by a pile of trash as tall as the hood of the car.

Welcome to Kamilo Beach.

Marcus tells me he is looking for recurrence—whether new materials have washed up on shore or whether the same trash washes in and out with the tides. Moore, in a previous outing here, placed marked caps at certain beach points and Marcus will try to find them. He's also trying to gauge the amount of trash by type and mass. According to the United Nations Environment Programme, on average 46,000 pieces of plastic litter are floating on every square mile of ocean. About 70 percent sinks. Marcus wants to know how much washes back up on shore. By the looks of it, tons do.

The tides are so strong here that a massive tree trunk, about twelve feet by twelve feet, is wedged into the sand not far away. Locals often come here for driftwood.

When you see it—the piles of trash washed ashore—it's shocking. Mounds of entangled fishnets are the first things I see. They are curiously stunning. Electric blues, neon greens, and bright yellows thread around brown seaweed and sand weeds pillowing together and creating a landscape of monster sea mash. From my notes: plastic oil containers, dish detergent bottles, a car mat, ropes of various sizes, buttons, plastic bottle caps, a size 8½ shoe, plastic butane lights, shards of glass, a green marble, plastic bags, plastic plates, a deflated volleyball, battery packs, a tire, water bottles, tin cans . . .

The distinguishable objects are one thing, but it's the millions of little plastic pieces that are awesome. They are multicolored chips that dot the entire beach, like sand.

The beaches and reefs from here to Kure Atoll, about fifteen hundred miles away and the westernmost point of the Hawaiian Islands, are the ultimate ciphers of garbage for the Eastern Garbage Patch and lend big clues about its effects on marine life and the ocean. Marcus examines plastics by type and chemical compound to figure effects.

While Kamilo Beach is a mess of trash, the water washing to shore is clear and looks clean enough for a swim. So I dive in. I can see clear through to the bottom. A crumpled can of orange soda rusts next to a rock; plastics hide in jetties; nets mask seashells. I decide not to swim too long.

It takes the better part of a day to gather enough evidence and samples to satisfy the Algalita team. We compare objects. The most intriguing: Jeff found an emergency transmitting device from a ship. It's like the "black box" on an airplane and is supposed to stay housed to a boat so it, or its wreckage, can be found and identified. He's eager to trace it and see what happened to the ship from which it came. Marcus found a toy soldier. I discovered a whiskey bottle from Japan, with whiskey still in it. And one of the other staff found a large bag of weed. We all agreed that was probably locally grown. For the record, we didn't smoke the weed or drink the whiskey.

We pile into the Trailblazer, and I have to think of it all the way back again: "Would you like to add insurance for $5.99 a day?"

THE WAVES ROLL far offshore, lumbering six at a time. The catamaran rises to each crest and then slaps twice—the fore of the pontoons first, then the midship where its twenty-five tons of weight is heaviest—and down the other side as we head out to sea. Ripples rustle up hoarse and sibilant. We are at the spot where it could be said, "the fierce old mother endlessly cries for her castaways." From this breakpoint the currents rush to Hilo's shore.

There won't be enough visible garbage here for me to see. The patch is elusive. It scales at different times and in different places. Even though we are ostensibly sailing on top of it, the matter here could be extremely diffuse.

Down below, Moore shows me his satellite gear, radar, and other technical equipment designed to track garbage particles. His mission this time across, his eighth, is to cover new territory and compare findings as well as concentrations during this time of year. Moore has never—well, no one has ever—sampled debris on a winter voyage.

During the winter months, currents tend to accumulate debris, which reach maximum concentrations in the spring, before summer current patterns disperse them again.

Moore's plan is to spend a month at sea investigating concentrations north of Hawaii and just east of the International Date Line. I wish I could go, but I'm relegated to joining the crew as they do drills, check equipment, and prep the vessel on the outskirts of the patch, not in the thick of it.

The expedition will also study deeper regions of the ocean, venturing into the "mixed layer" to see how much plastic is present near the limit where light penetrates and photosynthesis takes place.

Photosynthesis by ocean phytoplankton—or tiny drifting marine algae—is a major mechanism by which carbon is sequestered from the atmosphere. In fact, ocean phytoplankton conducts half of all photosynthesis on the planet. Land plants account for the other half. Ocean photosynthesis occurs at both the ocean surface as well as throughout the mixed layer—which extends down an average of 75 meters, or about 250 feet. Carbon dioxide dissolves into ocean water from the atmosphere and is "breathed-in" by these marine algae to grow their biomass. Now, for carbon sequestration to actually occur, these algae need to be eaten by marine zooplankton. Otherwise, they'll die and release their carbon back into the ocean or atmosphere through the decomposition process. If the algae are eaten, the carbon enters the zooplankton waste, which falls to depths below the ocean's mixing layer. Although down at these depths zooplankton carbon waste is eventually converted back into carbon dioxide, it is now in a very stable ocean layer known by oceanographers as a hundred-year time horizon. Here—which begins five hundred meters below the surface—water is unlikely to come into contact with the surface for a century or more.

You may be wondering what this has to do with ocean plastic. Simply put, if ocean debris exists not only on the sea surface, but throughout the mixing layer, it is undoubtedly acting as a sunlight filter, which is likely inhibiting ocean photosynthesis. If phytoplankton populations are unable to conduct photosynthesis, the chain of events that lead to zooplankton waste being trapped within that hundred-year time horizon will be severely hampered. The result will be more and more carbon dioxide in the atmosphere, which will only exacerbate the concentrations of greenhouse gases due to human emissions. In addition, there's some concern that the pieces of plastic covering the seafloor—which are estimated to number between 200,000 and 2 million pieces per square mile—may be creating an impermeable layer of debris that will eventually prevent carbon from seeping into the ocean sediments and being locked away for good.

So in effect, the extent of plastic's damage on the marine ecosystem could be far graver than merely pollution itself; it may upset Earth's natural balance and disrupt the huge role of the oceans in regulating the global carbon cycle. Plastics could, very simply, be preventing the oceans from hiding vast amounts of carbon, making the Earth heat more rapidly than it naturally would.

WE DROP SAIL, come about, and begin to follow the currents back to shore. We'd be moving just as debris would, except a hell of a lot faster. It takes California debris between four and five years to make its way out to the garbage patch, according to Moore's research. He says debris from Asia moves quicker; the stuff leaving Japan goes straight to the gyre in six months.

When we dock, I corral Joel Paschal, who is crewing with Moore, but who, until recently, worked with the National Oceanic and Atmospheric Administration (NOAA) on its research and cleanup effort of the Northwestern Hawaiian Islands. These are places like Midway and the French Frigate Shoals, which are largely unpopulated. But this isn't to say our refuse isn't there. Joel's job on NOAA's mission was to remove entanglement hazards, such as fishnets, from these protected areas (President George W. Bush in 2006 declared the Northwestern Hawaiian Islands a marine national monument) to ensure wildlife isn't disturbed.

Joel found a lot more than fishing nets. "There were all sorts of trash and plastic debris," he says. "Bottle caps, lighters, pieces of toys and parts of an airplane from World War II." He pauses. "Some of these islands have never been inhabited."

What struck him even more than the garbage were all of the dead birds.

More than 100,000 albatross and other seabirds are killed each year "inadvertently," according to the World Wildlife Fund. Many of these are caught up in long-line fishing nets along with more than 300,000 whales, dolphins, and porpoises and 250,000 sea turtles. The digestion of plastic debris kills so many albatross that they are considered an endangered species. In all, scientists have identified 267 types of marine species known to have been injured or killed by ocean debris.

Joel says, "It's tremendous to think about how remote you are. You're on an island far away from any people but you see our impact. Somehow the Earth is so . . . connected. You can go to the most remote island on Earth and there's garbage on it." Joel's not exaggerating. Even near the South Pole, plastic has been found in the guts of snow petrel chicks—one of only three bird species in the world that breed exclusively in Antarctica.

Marcus too spent time on Midway Island examining the effect of trash on albatross. "I found birds with syringes sticking out of their stomachs and toothbrushes caught in their mouths," Marcus says. He explains that albatross have two stomachs: one where they store food to feed their young and another where they digest food for themselves; plastic filled both. When he arrived on Midway he was shown a series of Laysan albatross skeletons. "Every single one had plastic in it. Every single one."

Marcus is clearly passionate about his work. It stems from a very emotional and wrought time in his life.

"I was in the first Gulf War. I was in the Marines. I was in a situation, and I was talking to another marine and I said, 'Man, if we ever get out of this let's do a Huck Finn and float down the Mississippi on a raft,'" he says. After the Marines, Marcus, now thirty-eight, embarked on his career in academia, earning his PhD in science education from the University of Southern California. His experience on Midway shone a light on the plastics issue, awakening him to just how massive the plastic trash problem is, and that idea of a rafting trip really never left his mind. So he decided to combine the two and create awareness about the plastics problem by embarking on a two-thousand-mile journey down the Mississippi River on a raft of plastic bottles.

"The Mississippi River drains 42 percent of the United States. And all up and down it I saw trash. It flows past New Orleans, where I'm from, and then out to sea. It took me five months to come down the Mississippi River. I couldn't come back and turn my back on all that trash. I couldn't just ignore this. It became a moral issue for me."

Marcus is now almost singularly focused on creating awareness. He teaches high school students about plastics and debris. He helps them build rafts as school projects. He has taken other rafting trips down the California coastline to create more awareness. And he is planning a big adventure across the Pacific, again on a raft made of plastic bottles. (You can also catch him giving "Commando" weather reports on the Weather Channel.)

Awareness is important. Some 8 million items of marine debris have been estimated to enter oceans and seas every day. Begun in 1986, the International Coastal Cleanup Day, organized by the Ocean Conservancy, has involved over 6 million volunteers in 100 countries who have scoured more than 200,000 miles of coastline and waterways, collecting more than 120 million pounds of marine litter. A majority of the debris found has been attributed to shoreline and recreational activities, such as beach picnicking and general littering.

The pathway from land to sea follows streams and rivers and underground waterways, as well as the wind. But it's the ocean currents that bring trash together in the Eastern Garbage Patch.

There are five oceans in the world and many seas, and their waters are continually on the move. These movements significantly influence the climate of the planet. Ocean currents flow around the globe in complicated patterns of warm and cold surface currents, eastern and western boundary currents, and the granddaddy current of them all, the deep waters that make up what is known as the global "conveyor belt."

Surface currents exist within the top four hundred meters of the ocean and make up 10 percent of all ocean water. The more well-known surface currents are the Gulf Stream in the Atlantic Ocean and the California Current in the Pacific Ocean. But each ocean has many. Surface currents flow in circular patterns, being driven mainly by atmospheric winds and the rotation of the Earth. In the Northern Hemisphere, surface currents flow clockwise; in the Southern Hemisphere, they flow counterclockwise. Warm surface currents flow from the tropics along eastern land boundaries to the higher latitudes; cold surface currents flow from temperate and polar zones along the western land boundaries toward the equator. This is part of the reason why water off the Atlantic coast of the United States is warmer than the water off the Pacific coast.

The deep water currents that drive the global conveyor belt generally move in opposite directions of the surface currents. The deep water currents that make up 90 percent of all ocean water are set in motion when warm surface waters move from the tropics toward the North Atlantic and become colder and saltier as they approach the pole. This cold, salty water is dense—dense enough to sink to the ocean floor, where it begins to move south, slower than molasses, along the Atlantic Ocean basin, eventually circling Antarctica, and then moving northward along the sea-floors of both the Indian Ocean (where it eventually resurfaces near the coastline of India) and the Pacific Ocean (where it eventually resurfaces in the North Pacific). It can take a thousand years or more for water from the North Atlantic to finally reach the North Pacific.

There are consistent currents that move throughout the oceans, and upon which oceanographers rely to predict marine and beach conditions, such as riptides, wave heights, and storms. Ocean current patterns can even help to better direct shipping traffic.

Take a piece of trash that has floated out of the mouth of San Francisco Bay, or has blown offshore in Malibu. You may see it get caught up in the California Current, which begins near British Columbia and flows south along the west coast of the United States toward Central America. From there, this piece of trash could get swept up in the North Equatorial Current, which would force it out toward Asia, where it might then swirl into the Kuroshio Current, which heads up the east coast of Japan before heading back east toward the United States. If it finds its way into a branch of the Kuroshio that graces the waters north of Hawaii, this piece of trash will be pushed into the large, eastward moving North Pacific Current, which then bifurcates into the Alaska Current to the north and the California Current to the south. This circle of currents comprises the North Pacific Gyre in the middle of which sits the Eastern Garbage Patch.

The oceans are patient. As mentioned already, deep water currents are estimated to take a thousand years or more before they move water from the North Atlantic to the North Pacific. But that's not to take anything away from the importance of the ocean's currents. Indeed, the benign purpose of many of the surface currents is to redistribute excess heat from the warm tropics to cooler areas at the higher latitudes in order to maintain livable conditions in areas like Canada, Western Europe, and the southern portion of South America. Other benefits occur as seasonal winds combine with the rotation of the Earth to drive warm, coastal waters out to sea. As a result, cold, nutrient-rich deep water wells up to take its place. This phenomenon, called upwelling, is particularly apparent along California's central coast. Upwelling not only stimulates an abundance of marine life, it makes fisheries more productive because of the nutrient release. You'll notice it occurring because it produces the foggy summertime conditions that inspire the term *June gloom.*

Over the past several decades, our knowledge of ocean currents has grown exponentially. As computer modeling improves, we're understanding more and more about how currents are influenced by wind, water temperature, salinity, the Earth's rotation, and ocean-floor topography, as well as how currents influence local, regional, and global climate; plant and animal productivity; rain patterns; and carbon dioxide levels in the atmosphere. So far, we really know very little about how currents may be affected by new particles entering their stream, now that we've inserted billions of tons of trash into the mix.

If the calculations are accurate, there are about a thousand pieces of trash at sea for each of the 6.7 billion people in the world.

Moore says we have to stop. We have to relieve the oceans of their burden and begin to clean up our acts onshore because, as his work exemplifies, it ends up offshore too.

"There is no such thing as throwing anything away," Moore says.

And there is no way to clean up the mass of garbage in the oceans, not even if we restricted our cleanup to the Eastern Garbage Patch. The oceans are too vast.

The Pacific Ocean, named by Portuguese explorer Ferdinand Magellan ("peaceful sea" in Latin), is the largest body of water on the planet. It is more than 60 million square miles, an area large enough to house all the land in the world, and it's deep—35,840 feet at its deepest point. You can't vacuum-clean it.

"There aren't enough vessels in all the navies of the world to embark on such a mission," Moore says. "It is the largest habitat on Earth. Its average depth is two miles and in many cases we don't know how deep it is. Even if we could clean this one, there are three other gyres like it in the Indian Ocean, the Atlantic, and in the South Pacific."

If you do the math, there are 3 *trillion* pieces of plastic debris floating in the Pacific Ocean alone.

Moore says the solution isn't cleaning up our past but cleaning up our future.

"Waste is an antiquated concept. The only reason to throw anything away is because it's difficult to reuse. We don't have to waste anything, we just have to start making things that are easy to reuse or recycle," he says.

Until then, we have to control the litter and the sewage that washes offshore.

California has begun an initiative to start cleaning up the Los Angeles River basin. And other basin areas are beginning similar programs. They net-trap debris in the water, filter runoffs, collect trash on banks, and host programs to educate the public. Schools and volunteer cleanup campaigns have also been tapped to regularly collect trash. The Great Los Angeles River Cleanup, La Gran Limpieza, collects twenty-five tons of trash in one day, which includes mostly plastic bags, candy wrappers, chip bags, and polystyrene cups.

In Japan a company is exploring ways to wring out the petroleum base from saltwater and create an energy source from it. Petroleum is a base ingredient in plastics, as you may recall.

But Moore isn't bullish on that happening in a big way or on it being a solution. A more important and tactical solution is raising people's awareness of all this so they can take steps to mitigate their own trash burden. "We need to show them what's out there," he says.

You Are Here's expert researcher, Dr. Howell, agrees. She says that people can take some simple steps to reduce the chances of their waste ending up in the ocean. After all, 80 percent of marine debris originates on land:

- When you visit parks and beaches, keep an eye on food wrappers, plastic sacks, paper plates, drink containers, and other lightweight disposable items that can easily take to flight with even the slightest gust of wind. Make sure you either carry these items out with you, or dispose of them properly in trash bins. An even better idea for preventing litter is to pack your food and drink in reusable containers and reusable canvas bags. If your outing involves sand toys, plastic Frisbees, goggles, flippers, beach balls, or other recreational goods, make sure you retrieve them before you leave. Regardless of whether these items seem like "litter," they'll eventually be swept out to sea and become threats to the health of marine animals.

- Around town, you can help prevent marine debris by keeping trash out of streets, gutters, walkways, parking lots, lawns, sidewalks, and pretty much everywhere else litter might accumulate. This litter usually finds its way into a local storm drain via rain or wind and then—regardless of how close you live to the nearest coast—it's eventually discharged into an ocean, bay, sea, or gulf.

- Even at home, your actions may help reduce the volume of global marine debris. For example, think twice before using toilets as garbage bins for trash and personal care products. Although sewage treatment plants are designed to filter out bulky items, it's not uncommon for untreated sewage—especially if it's combined with storm water—to be discharged directly into a nearby river or ocean. So remember to use the trash can instead of the toilet for things (use your imagination here) made from latex, plastic, metal, and silicone.

- Never dump kitchen grease down drains or garbage disposals. Grease is the most common cause of sewage-system blockages, which can result in overflows or spills into oceans and other waterways.

- In addition to being more conscious of your own actions, consider joining a grassroots or national campaign to beautify a local park, neighborhood, beach, or riverbank.

Students and artists in Hilo have put together a really creative solution: every year they stage a trash-art fair. They make pieces of art out of the vast and diverse volume of trash that washes ashore on the island, and sell them. Their solution not only removes waste, it creates awareness and raises funds that go toward future cleanup efforts.

The fair takes place near the spot where I am standing as the crew of the *Alguita* stocks the boat with unripe fruits and vegetables for their voyage. They store meat and milk and bread and eggs. They need supplies sufficient for the eight people who will be aboard for a month. I wish them good-bye, and just before sunset on a midwinter's day, they set sail.

They post a blog every day from sea:

Our first three trawls, to the naked eye, yielded scattered pieces of plastic, a few visible nurdles, and a host of colorful organisms—numerous Vellelidea "blue buttons," copepods, salps, Portuguese man-of-wars, and other minuscule creatures. We won't know for certain how much plastic these samples contain until we bring them back to our lab.

It really is difficult to comprehend the vastness of this phenomenon. There is still a common public misconception that the gyre is a "place," a detectable spot, when rather it is an enormous, extremely diffuse region. . . . Being out here, seeing nothing but blue horizons day after day certainly helps.

We began the day with some early morning sampling—Marcus, Joel, and the captain pulled up the Manta Trawl first thing, to find the by-now-predictable tangle of tiny life forms interspersed with plastic particles.

The ship's blogs, after more than a week out to sea, prove the importance of the expedition and its mission to chronicle debris findings at every latitude.

Plastics parting their way in swarms through waves and settling to the ocean's depths. Or drips of saline congealing and dappling in glutinous pearls of polymers on its surface. The largesse of our human footprint in this way can be found thousands of miles out to sea, walking across the surface of the water, or standing ever taller on its floor.

Still, although between 60 and 80 percent of marine debris is plastic, other types of refuse also swim out of our discards.

From hypodermic needles and radioactive military waste, to toy soldiers and birthday balloons, marine pollution is diverse. It ranges from microscopic to titanic; chemical to physical; terrifying to benign. Its sources include (but aren't limited to) oil spills, air pollution, factory effluent, agrochemical runoff, cruise liner dumping, and sewage discharge. Ocean contaminants involving heavy metals or fat-soluble chemicals like mercury, dioxins, or PCBs tend to bioaccumulate, or build up, in the bodies of marine organisms as they're progressively eaten by larger and larger predators. For this reason, the Food and Drug Administration issues warnings about eating too much seafood, as such toxins can easily assimilate into and injure our bodies as well.

Mercury buildup in seafood especially makes the news, and the FDA explains:

> Mercury occurs naturally in the environment and can also be released into the air through industrial pollution. Mercury falls from the air and can accumulate in streams and oceans and is turned into methylmercury in the water. It is this type of mercury that can be harmful to your unborn baby and young child. Fish absorb the methylmercury as they feed in these waters and so it builds up in them. It builds up more in some types of fish and shellfish than others, depending on what the fish eat, which is why the levels vary.

It's hard to overemphasize the sensitivity of aquatic species. And for this reason, every effort should be made to avoid dumping toxic substances—including prescription drugs—down the toilet or drain. Where possible, try using plant-based and toxin-free products for gardening, laundry, housecleaning, and dishwashing.

You can actually trace waste material from a garbage disposal in a home in Boise all the way out to the Pacific, even to the Eastern Garbage Patch itself. Start at the mouth of the Columbia River on the border of Oregon and Washington, and work backward. Navigating through its tributaries you land in Idaho. From underground storm water systems, drains, and sewers you can trace materials to a garbage disposal inside a home.

Water seeks out and finds water, whether salt or fresh. Indeed, there is no such a thing as "fresh" water, nor has there ever been. You'll see what I mean by that in the next chapter, as I follow the water trails of our existence from the seas into our homes and into our bodies.

NINE

The Greatest Problem No One Has Heard About

The Great Lakes, Duluth, Minnesota

The human body can survive for only about seventy-two hours without water. Obviously water is vital to our existence. And with the population growing and freshwater supplies drying up, water is becoming increasingly scarce. By the year 2025 the world will need to find a way to increase its freshwater supply by 20 percent in order to meet current demand. And that ain't easy; you can't just make freshwater.

This leaves just one other option: we have to mind what we have. The amount of water we waste is profligate. Each of us over the course of a year uses about 20,000 gallons of water—to drink, bathe, wash things, flush the toilet. We waste 10 percent of that, or 2,000 gallons, just by letting the taps run, the drips amount, and the leaks continue.

At the group level it's even worse: we let freshwater drain out to sea. We pollute groundwater, streams, and lakes. We let our infrastructure rot and our water systems break down.

According to some estimates, it's going to cost more than $3 trillion to upgrade the water systems in the United States. Worldwide, the costs will soar to more than $20 trillion. It's a huge issue that has massive implications for the population. Some people, particularly government officials, have known this for a long time. Some ask for money to upgrade and repair the systems. Others make a fuss about it. But requests are dismissed and the fuss fades. Other issues make the headlines; other topics make for easier policy. Proposals end up going nowhere.

Look at the water situation in the United States, for example. The Clean Water Act was passed in 1972. The purpose of the act is to protect the nation's "navigable" surface waters from harmful contaminants.

With the exception of some minor changes here and there—most recently the definition of "navigable waters" was altered in such a way that fewer bodies of water are protected—the Clean Water Act hasn't been updated or improved upon. Why? Because water is the dirty secret around the world that no one wants to address; because the problem is so scary, complex, politicized, expensive, and potentially lethal that it's easier to just leave it up to other people—or other generations—to deal with. Unfortunately we're running out of time.

Two highly respected PhD water experts, Bill Jury and Henry Vaux Jr., recently issued this warning: "Without immediate action and global cooperation, a water supply and water pollution crisis of unimaginable dimensions will confront humanity, limiting food production, drinking water access, and the survival of innumerable species on the planet." And these scientists are not prone to hyperbole.

More than 1.1 billion people do not have access to safe water, and that includes hundreds of millions of children. It doesn't mean there isn't enough water for these children. It just means that it is beyond their reach. We can get it to them. We can do something about this.

Five million people die unnecessarily each year because of illness related to lack of potable water. Half of them are children under the age of five. To bring it home, think about this: one child dies from lack of clean water every twelve seconds.

Please pause and read that last sentence as more than facts and words and numbers on a page. This is where the preservation of natural resources begs serious attention and should incite compassion and care in each and every one of us. Most residents of the developing world get by on a little more than five gallons of water per day; the average global citizen uses about thirteen gallons per day; all the while, water use in Western Europe and the United States ranges between 50 and 170 gallons per person per day. Think we can get by on using a little less and putting a little more into the hands of people who need it? Part of the travesty is that the world's poorest people typically pay five to ten times more per unit of water than do people with access to piped water. Water extraction and storage inefficiencies make water more expensive. These people often inhabit areas with little or no existing water supplies. It must be transported. Developed countries have efficient water systems that allow the relative ease and economy of access to freshwater. But underdeveloped countries do not have such systems in place and therefore freshwater becomes a much higher-priced commodity.

The heart of the solution to the water issue lies in investments to create better systems and improve quality. More efficient systems would also mitigate water leakage and runoff, giving us more water from existing sources. Investments would allow us to use less without our really noticing any difference.

John Dickerson is a water investor. He has followed the freshwater issue for years, operating investment firm Summit Global Management, which invests in utilities, private water companies, and big multinational water concerns. In other words, he invests in the business of water. A former spy with the Central Intelligence Agency, Dickinson keeps tabs on the water industry around the world. Close tabs.

John explains that the water business is growing because as supply shrinks, demand rises, and that means prices go up in all sorts of ways. Water, as an investment, has outperformed the stock market for at least the past twenty-five years.

I first met John about a decade ago while I was writing a story about water, and I've kept in touch with him whenever I've had a water-related question.

John operates out of a nondescript office building just outside San Diego. Inside his office he keeps lots of reference materials and books. John is chock-full of facts:

- Developed countries are struggling to maintain their aging infrastructure. The United States alone has 700,000 miles of drinking-water pipe, some more than a hundred years old.

- Available freshwater is less than ½ of 1 percent of all the water on Earth.

- Eighty percent of the global population relies on groundwater supplies that are dangerously depleted, if not exhausted, as they are mined beyond natural replenishment.

- Pollution and climate change further exacerbate supply shortages, damaging vulnerable resources and causing drought and desertification at an alarming rate.

- Per capita water consumption has roughly doubled in the last century, a rate that will accelerate as more economies industrialize and populations become more urban.

These facts tie together many of the issues that have been raised already in this book: urbanization, pollution, lack of awareness. When we connect leaving the lights on to energy shortages and pollution rising, when we connect our trash to landfills and disease, and when we connect the things that we buy and use to the places from which they came, we more fully understand the importance of being mindful. Ignoring the causes and effects of what we do and how our actions impact the planet—as we should all understand by now—will eventually come back to endanger us. And water is the issue in need of the most immediate attention and awareness. Without water, all living things on the planet die.

A big contributor to the water problem facing us is global warming. Climate change is expected to account for about 20 percent of the global increase in water scarcity in the coming years. If we don't do anything about global warming, the severe droughts that currently occur about once every fifty years will be occurring every other year by 2100. If you thought sea level rise was a severe consequence of melting Arctic ice, take stock of this: almost half of the world's population relies on water flowing from Himalayan glaciers that are projected to melt away within the next fifty years.

With water such an important part of our survival, and with all the money and issues that go along with trying to preserve its quality and conserve its supplies, what I want to know is why water can take such a backseat to any issue put in front of us today, including energy.

"I can't explain it," John says. "Water is an everyday thing. It's all around us. People don't see it as a problem, until they start looking at it on a global basis. And that is how we have to start thinking about water. Maybe then perceptions will change. It's a bigger issue than energy; there is no substitute for it at any price . . . and some countries have more of it than others, so you can easily see how conflicts could arise; wars."

It may sound bizarre that a military battle would be waged over something we take for granted such as water. When you add up the facts that China has 22 percent of the world's population but only 8 percent of the water supply, and that 60 percent of the world's freshwater is found in just nine countries (there are 195 countries in the world, if you include Taiwan), you begin to realize how wars could be waged over water resources.

Still, it shouldn't have to come to that. There is enough water in the world to go around. It just needs to be cared for and allocated properly.

Filling your glass three-quarters full from the tap isn't what what we are talking about here; drinking isn't the biggest water consumer. What sucks up the most water is agriculture.

Today, agriculture accounts for about 70 percent of all water use in the world, and up to 95 percent in several developing countries. Here's why: to produce enough food to satisfy the average person's daily diet requires between five hundred and eight hundred gallons of water. You need water for seeds, fruits, vegetables, and for cattle, chicks, pigs, and everything else we live on. Then there's the processing, manufacturing, packaging, delivery, and storage of these goods. Water, water, everywhere, at every step of the way.

Water supplies are estimated to be 17 percent short of the amount necessary to feed the global population by 2020. Three of the world's largest grain producers—China, India, and the United States—face the most severe water supply shortages. The nations that rely on these three for grains, which is most of the world, in turn face a food crisis. Grains are a basic element in the food supply chain, for us as well as for the animals on which we feed. What do you think cattle, chickens, and pigs eat?

The amount of water that goes into producing grains, meat, and other commodities is called "virtual water." For example, it takes about 120 gallons of water to grow a pound of wheat. So the virtual water of this pound of wheat is 120 gallons. For a pound of meat, the virtual water is thirty times higher. It's important to know this because virtual water accounts for 15 percent of global water consumption, and is where major water savings can accrue. When food crops and commodities are traded, there's a virtual flow of water from the producing country to the importing country. Virtual water saves almost 5 percent of the water used annually in global agricultural production because places with less water gain access to foods with high water requirements by importing them from areas with high rainfall or substantial water supplies. These water-scarce regions can then use their own water resources more efficiently for other purposes—and create savings. This means areas of southern China, which have more water and are better equipped to grow certain water-intensive agricultural products, can send these products to the north, which frees up water supplies in northern China for other uses. It's about being smart with water uses.

There's even a movement afoot for manufacturers to begin labeling the virtual water contained within their products, allowing us to see clearly how much water it takes to make, say, a dozen hot dogs, or a carton of ice cream. The World Water Council, an international organization that seeks to create awareness about water, encourages people to eat lower on the food chain—fruits, vegetables, legumes—in an effort to conserve virtual water. Meat requires far more water to bring to plate than greens. It takes 370 gallons of water to produce one pound of rice, whereas it takes a whopping 3,434 gallons of water to produce one pound of beef.

"It is clear that moderating our diets especially in the developed world could make much more water available for other purposes," the World Water Council says in regard to virtual water.

I could live with making better choices in my diet if it would allow more water to be used for foods that are shipped to developing countries, allowing those countries, in turn, to use their scarce water resources for things they desperately need—like drinking water. I'd swap a burger for a salad every once in a while. This is how water-saving actions here affect people, places, and things "there."

We, of course, can't really see virtual water; it's lapped up by all the things we consume. What I can see, however, is one of the largest sources of freshwater on the planet—and it isn't all that far away.

The Great Lakes–St. Lawrence River system, which runs from the Midwest of the United States to the Atlantic Ocean, is the largest surface freshwater system in the world. The Great Lakes comprise Lake Erie, Lake Huron, Lake Michigan, Lake Ontario, and Lake Superior.

They are an amazing natural wonder, and they are right here in our backyard. Not that we appreciate the lakes all that much. I, for one, really never thought about how valuable a resource the Great Lakes are to the country. But they are, particularly now, as water becomes such a valuable commodity.

Countries in Asia have tried to buy water from the Great Lakes. In fact, a decade ago a Canadian company was issued a permit from the Ontario government to withdraw 158 million gallons of Lake Superior water per year in order to ship it by tanker to Asia. Public outcry and protests from members of the

US Congress eventually forced the company to abandon the plan before it began. However, schemes to export the Great Lakes' water still abound.

The Great Lakes are divided up between Canada and the United States. The Canadian province of Ontario along with the US states of Minnesota, Wisconsin, Michigan, Illinois, Indiana, Ohio, Pennsylvania, and New York border the lakes and therefore have ownership of at least part of the water system. The eight Great Lakes Governors have all signed an agreement that would prevent most future diversion proposals and all long-distance ones, but there is still haggling among them about water withdrawals. How much can one city, for example, divert to its suburbs? You can see how complexities arise.

Beyond self-regulation, military plans have been drawn up to protect the Great Lakes. In a fantastic article in *Harper's* magazine in September 2007, far-out schemes of freshwater diversions to the United States were cited: diverting Canadian rivers to run southward rather than northward; fjords in British Columbia dammed; tankers filled with freshwater at Canadian ports and sent south to Los Angeles. All considerations because of the ever-increasing US water shortage.

The US government projects that at least thirty-six states will face water shortages within five years because of a combination of rising temperatures, drought, population growth, urban sprawl, waste, and excess. Indeed, during the 2008 presidential election primaries New Mexico Governor Bill Richardson "created a stir," as newspapers called it, when he was campaigning in water-hungry Las Vegas and called for a national water policy. His remark that states like Wisconsin are "awash in water" set off alarm bells.

"No one has seriously proposed that parched western states sip from the Midwest. And Mr. Richardson's office swiftly declared he had no such intention. But his remark tapped a growing sensitivity here over the Great Lakes and has given new urgency to a regional initiative to protect them from outsiders," the *Christian Science Monitor* wrote at the time.

Who knows, maybe water will have to be shipped. In the summer of 2007, Georgia, Florida, California, Texas, Arizona, Nevada, and other states faced a water shortage crisis. Georgia had to beg the federal government for emergency assistance because water supplies got so low that state officials predicted it would run out of water within three months.

And anyone who watched the wildfires along the California coast and elsewhere in the summer of 2008 knows the consequences of drought and dry lands.

As it stands, the laws governing water ownership are weird. Some water is regulated by the federal government, some by the states, and some by who got to it first.

In many states in the western United States, for example, water is divvied by "prior appropriation." That means the first person who claimed access to the water gets to use as much as he or she needs, then the next person in line is allocated as much as they need, then the next person, and so on. Water titles are separate from land titles.

John Dickerson says many of the shortage issues could be easily wiped away just by nixing this covenant. "You have farmers like [some] alfalfa growers that needlessly grow excess crop and then plough it back under just so they can maintain their water titles. The enormous amount of water they use could easily be enough for all the people in California. But those farmers know that water title will be worth zillions one day so they keep artificially farming, and the government doesn't do anything about it," he alleges.

People with rights and access to water are quickly realizing they own a valuable commodity. The Great Lakes' owners understand this and don't want to part with any of their water either.

"There have been a number of ideas to transfer water out of the Great Lakes. Some years back they wanted to use it to raise the water levels on the Mississippi River. There was talk of a pipeline from Wyoming to Montana with some thoughts on replenishing the aquifers beneath the Great Plains states," Dave Naftzger, executive director of the Council of Great Lakes Governors, tells me. "The governors have gone to great lengths to ensure strong protections of the Great Lakes water as a resource and to power the regional economy."

THERE IS A pathway that begins just over Frontage Road in Duluth, Minnesota, and extends past docks, bridges, channels, and beaches. It winds along the banks of Lake Superior. There are shops, restaurants, statues, and monuments all in some way related to the lake and its freshwater. Where the shore meets a small shipping channel, I stop and read a sign that has been staked into the ground:

> Lake Superior is 31,280 square miles, equal in area to Massachusetts, Connecticut, Rhode Island, Vermont, and New Hampshire—combined. It stretches 350 miles east to west and 160 miles north to south. It is the deepest of the Great Lakes; it is over one-quarter mile deep. It was filled with glacial meltwater 10,000 years ago. It holds three quadrillion gallons of water.

—and this is the part that stunned me—

> That's enough water to flood all of North and South America with a foot of water. If it were emptied it would not be refilled by its natural flow until 2179. Over two hundred rivers flow

into Lake Superior. When you take into account all five Great Lakes, their size eclipses that of the United Kingdom.

This vast body of water sits mightily atop our border. It is 2,300 miles from the Atlantic. It would take the average vessel about a week to travel that distance. Here, in the middle of the country, you can find ships from Asia, Europe, and the Middle East—in the middle of our country.

It's phenomenal that we have this resource and that it allows us to connect from the heartland to the rest of the world.

Yet with all the bluster of military protection and feuds over withdrawals, ownership, and the like, the Great Lakes are suffering. Pollution is seriously affecting water quality: ninety cities dump sewage into them. More than 30 million people, 10 percent of the entire US population, rely on the Great Lakes for freshwater.

As well, temperature rise is causing extreme evaporation. Lake Superior is at its lowest point in eighty years, and the other lakes are seeing water levels dip too. Beyond municipal use, there are shipping ramifications: commerce halts when water levels run low and ships run aground.

The average water temperature of the lakes has risen more than four degrees Fahrenheit since 1980. That's four times the average increase in global atmospheric temperature—which is a lot. When rises like this occur, water begins to evaporate at a faster rate. Global warming, as noted, is quickly drying up our freshwater supplies right here, right now.

As I'm milling around the Holiday Inn lobby in Duluth eating the home-baked chocolate chip cookies one of the staff made and reading a copy of the local shipping news, I strike up a conversation with Brian Knutson, a local Minnesotan who grew up around the Great Lakes. "We really don't think about not having freshwater," he says. "I'm so used to having so much water, growing up in Minnesota, it's strange to me that people have to think about it." Then he mentions news of a conservation conference that was recently held in town, and the awareness that it provoked: "Our water is evaporating, you know," he says, in that delicate way people do when they approach a sensitive subject, something that evokes awe, shock, or befuddlement. He shakes his head. He mentions meeting a woman from Texas who was complaining about the low water levels in her local reservoir. "She said it was down to 2 percent. I can't imagine relying on a reservoir for water," he says. It's evident that perhaps one day he too might have to give water use a whole lot more thought.

The shipping news I'm reading lists ships by name and describes their cargo: "*Philip R. Clarke* Expected to arrive Duluth bet 21:00 and 22:00 for Cutler Stone to discharge coal . . ."; "*St. Clair* Expected to arrive Duluth at 5:00 for Midwest Energy to load coal . . ."; "*Mesabi Miner* Expected to arrive Duluth bet 18:00 and 20:00 for Midwest Energy to load coal . . ."; and so on. Most of the sixteen ships expected to come and go that week were carrying cargoes of coal. Ironic, I think, that much of the cause of evaporation of the Great Lakes sails right on top of them.

When I draw the drapes to peer out the window of my hotel room, there's a ship staring me in the face. It's that close. I couldn't reach out and touch it, but I could certainly hit it if I chucked a baseball. Beyond, there's open water as far as the eye can see. If you were transplanted here with a blindfold on and then, blindfold off, told to say where you are, you'd probably guess some seaport. Lake Superior is actually bigger than some seas.

Duluth has a system of skywalks so that you barely have to walk outside to get to the shore. Because of the wind that whips up off the water and the cold winter temperatures, I gather the city figured it more convenient to build walkways above than to bundle up every time someone wanted to step outside.

Outside has the makings of small-town America: diners, a few clothing stores à la JC Penney. No big high rises. Nothing too new-looking or fancy. Slightly ramshackle, you might say. Duluth is not a sprawling metropolis. It's Mayberry with an edge. Besides the port, its biggest claim to fame may be that it's the birthplace of Bob Dylan.

Out the second-floor elevator and past the Subway restaurant, I head through the skywalk. It reminds me of walking down the corridors in junior high school when everybody else was in class. The corridors look and feel the same: cold gray and blue colors. I pass a curling gym where some kids are practicing sweeping in front of the disc another has "hurled" just like they do in shuffle-board—not something you see every day in Los Angeles.

I step outside and breeze past the visitors' center, the coast guard ship, and arrive at the tanker that was staring at me. Several stories high, it forces your head back when you look at it from ground level. It's docked so close to town because all the ships dock close to town. Railroad tracks run alongside the shore so tankers can be loaded and unloaded, making transport easy.

The busiest times for shipping on the Great Lakes are in August and September when Midwest grain season is over and the harvests are ready to be shipped abroad. Farmers load up trucks that in turn load up trains that in turn load up ships on the Great Lakes. From the ships, these crops are distributed all over the world. The biggest farm exports of the United States are corn, soybeans, cotton, and wheat. Conversely, ships bring to Duluth cargo containers of all sorts that are eventually loaded onto trucks, trains, ships, and planes and then transported throughout the country. Less than two hundred miles from Duluth is the Mississippi River, which makes its way to the South.

So there you have it: the vortex of our commerce and agricultural export system sitting on top of the biggest water source in the country.

I want to drink it. Down the bank I go, onto the rocks and sand. Apparently I am not the only one who has hopped down to the water's edge (even if I am the only person in sight on the entire shoreline now). Ducks are there too, and they spot me. Hundreds of them. They quack. I figure I have disturbed them and they will glide along away from me, perhaps a little pissed off that I have trampled on their turf. Nope. They paddle right at me. I dip my water bottle in the water and fill it. This doesn't put them off either. They form a line, like some redcoat revolutionary-times military formation. The leader, I swear, is staring into the whites of my eyes. I back up on shore. They are moving fast. The first battalion lands and breaks directly into a sprint right at me. What the f—? Ducks. Mallards. These birds aren't supposed to attack. But quack, quack, they do—and give chase.

I am now very thankful that I am the only one on the shoreline, otherwise I would be in the most embarrassing position of being seen fleeing the lake holding a water bottle and being chased by a flock of angry ducks.

Safely back on the strand above the rocks, the ducks waddle around. I sip cold, freshwater right from America's largest source.

ALL AROUND THE Great Lakes there are vending machines selling Dasani bottled water. Dasani is made by Coca-Cola. Coke says it gets its water for Dasani from public water sources wherever it's bottled. That could be Minnesota, or it could be its headquarters in Atlanta. Curious next to that large source of freshwater is bottled water that could be from Georgia, which doesn't have water to spare. Interestingly, 25 percent of all bottled water is from municipal water sources—tap water—not from sparkly mountaintops or special springs as some advertisements would have you think.

Tap water itself siphons only about 13 to 14 percent of total freshwater use, and only about 0.5 percent of that is used for drinking.

The controversy over bottled water use that has been raised is more about the excess use and waste of bottles than the water. It's what we do with water that is the crime.

Just about the exact same amount of water exists on the planet now as when dinosaurs roamed. We just use a hell of a lot more of it now. And something interesting to note is that we use the same water that the dinosaurs used.

Water is made up of molecules, two hydrogen atoms and one oxygen atom; it's H_2O. Throughout the "water cycle," water molecules bounce around, but they don't really disappear.

"Water at the bottom of Lake Superior may eventually fall as rain in Massachusetts. Runoff from the Massachusetts rain may drain into the Atlantic Ocean and circulate northeastward toward Iceland, destined to become part of a floe of sea ice, or, after evaporation to the atmosphere and precipitation as snow, become part of a glacier. Water molecules can take an immense variety of routes and branching trails that lead them again and again through the three phases: ice, liquid water, and water vapor. For instance, the water molecules that once fell 100 years ago as rain on your great-grandparents' farmhouse in Iowa might now be falling as snow on your driveway in California," NASA says.

The water cycle works like this: liquid water becomes water vapor (a gas) when moisture evaporates from oceans and lakes. Moisture escapes ice and snow in much the same way. Through a process called sublimation, water can go directly from the solid phase to the gas phase. Plants release moisture as well through a process called transpiration: when they take water in through their roots and deliver it to their tissues, some moisture is released through the leaves in order to keep the plant from being cooked by the sun.

The oceans are the primary drivers in releasing moisture into the atmosphere, while plants contribute only 10 percent. But plants can still make a lot of water. A one-acre cornfield can transpire 4,000 gallons of water every day.

After the moisture is released into the lower atmosphere, winds carry it upward high into the atmosphere where the air is cooler. Cool air can't retain as much moisture as warm air—as anyone who's been outside on a hot and humid day can confirm. As a result, the water vapor in the air condenses into cloud droplets, which fall to earth in either the liquid (rain) or solid, (snow, sleet, freezing rain, hail) phase.

When precipitation falls, some seeps deep into the ground and forms groundwater, some runs into rivers and streams, some enters back into the oceans and lakes—and then the cycle continues.

Of course, we and other forms of life interrupt this cycle and grab our share of water. Over the course of a year, the amount of water that is cycled through the atmosphere is enough to cover the Earth's surface to a depth of 97 centimeters, or more than three feet. Our use is but a fraction of that.

Precipitation and evaporation usually cancel out each other so water levels remain constant. But over the last hundred years, a shift has occurred. On land, precipitation now exceeds evaporation; over oceans, evaporation now exceeds precipitation. More rain and snow occur over land than seas, in other words.

This phenomenon would leave our oceans empty if runoff from land areas didn't send more into them. Now, however, too much water is being sent their way.

As global warming raises ocean temperatures, something called thermal expansion is occurring and the same quantity of water is taking up more volume. Thermal expansion is the concept in physics whereby things expand when heated. For example, water heated from 90 degrees F to 140 degrees F in a forty-gallon hot water heater will expand by almost one-half gallon. This is because when water is heated, its density decreases and its volume grows.

But the Earth's water cycle doesn't compensate for this by evaporating more moisture into the atmosphere. What happens? Sea levels rise. About half of the sea-level rise that has occurred over the past fifteen years is due to thermal expansion. The other half is due to the melting of ice sheets and glaciers. Both are caused by warming global temperatures.

We've seen the full-circle effect of that warming, how we are connected to it, and the causes of it, from our energy use to our trash. Higher temperatures on land mean more severe droughts. On seas, higher temperatures mean more water, but that water isn't potable, or "freshwater," that can be used for drinking, bathing, washing, or for producing foods or commodities. Too much salt. Logically, we need to switch things around and capture a bigger amount of water than what exists only on land.

People are working on it.

Experts and engineers say the desalination of seawater is the likely solution to make more freshwater available. But desalination is expensive. It can cost over a thousand dollars per acre-foot (roughly 325,850 gallons) to desalinate seawater as compared to about two hundred dollars per acre-foot for water from normal supply sources. Some inventors have designed technologies that halve that cost, but even then it's still double the cost of normal supply. The sheer volume of water we use makes desalination a very, very expensive proposition.

Another way to expand the freshwater supply is to reuse it. This would mean recapturing and recycling wastewater back into potable water. "Toilet to tap" processes are more economical than desalination, but perceptions are big obstacles. It's hard to trust and believe that the water sitting in a glass in front of you waiting for you to drink and that came from a sewer is clean and safe.

Toney Orange County, California, which sits between San Diego and Los Angeles, is rolling out one such pioneering program. It invested half a billion dollars into a facility that can turn 130 million gallons of treated sewage into drinking water every day.

The facility uses sewage that would have been otherwise discharged into the sea, and filters it.

An article in the *Los Angeles Times* describes how it's done:

The effluent is first pumped into the reclamation plant from the sanitation district's sewage treatment facility next door. The brackish water, which smells of deodorizer, flows into twenty-six holding basins equipped with 270 million micro-filters—thin straws of porous material with holes no bigger than three-hundredths the thickness of a human hair. From there, the water is forced under high pressure through a series of thin plastic membranes housed in rows of white cylinders. Next, it is dosed with hydrogen peroxide and bombarded with ultraviolet light to neutralize any remaining contaminants. At this point, the water is free of bacteria, viruses, carcinogens, hormones, chemicals, toxic heavy metals, fertilizers, pesticides and dissolved pharmaceuticals. Though it is good enough to drink, the scrubbing isn't finished. Once the state approves, up to 70 million gallons of treated water a day will be pumped into the county's giant underground aquifer. It will be cleansed further as it percolates through the earth to depths up to one thousand feet.

Sounds delicious. But at $550 per acre-foot, the recycled water is slightly more expensive than supplies brought in from Northern California. Still, it's a beginning.

Another way to virtually expand the water supply without the muck is to deal with the high levels of inefficiencies in freshwater distribution and delivery systems. It's estimated that many water distribution systems are losing up to 30 percent of the water captured due to cracked and/or leaking infrastructure alone.

"We let water wash right out to sea. How about instead of investing billions in desalination projects we turn the storm sewers east?" John Dickerson says. As they stand, storm sewers in California face west toward the ocean. Reversing their flow would allow water to be captured.

John also points out another logical and relatively inexpensive solution: due to the fact that precipitation is seasonal and longer-term weather patterns (El Niño, droughts, etc.) tend to be cyclical (as we learned way back in the first chapter in the discussion of teleconnections), capturing water during wet times to be used during later dry periods is an ideal solution. The methods John speaks of don't require new construction or technologies; they are methods we are already using—just not to their best potentials.

When it comes to farming, more efficient irrigation and slow-release, nutrient-rich fertilizers could enhance water productivity. Virtual water utility, of course, could also come into play.

Then there's us and how we use water.

The easiest thing to do without having to think much about how you're using water is to install water-saving devices. Low-flow toilets, low-flow faucet aerators, and low-flow showerheads are not expensive and could save a family of four more than

35,000 gallons of water per year—which is nearly three months' worth of indoor water use.

You've heard it time and again, but turning the water off while you brush your teeth really can make a difference—1,400 gallons per year.

Put a few drops of food coloring in the tank of your toilet before you go to bed. Don't use or flush the toilet all night. If the water in the bowl has changed color by morning, you have a leak. Repair the leak and you'll save up to 500 gallons of water you've been wasting every day since the leak began—you'll probably see a sizable difference on your water bill as well.

If you have a dishwasher, use it. Run full loads. And don't rinse your dishes before you load it up. Dishwashers use less than half the water used when washing dishes by hand. You'll save time, energy—after all, it takes energy to heat all the hot water you use—and about eleven gallons of water per load. If you do four loads per week, you'll save almost 2,300 gallons per year.

If you're a bath taker, consider becoming a shower taker instead. You could save about 12,000 gallons of water per year. Filling the tub uses three times more water than a ten-minute shower.

If you live in a dry region of the country, cut your outdoor water use by replacing your unquenchable lawn with drought-tolerant plants. If you're not keen on the native garden look, try planting some shade trees to reduce transpiration and the need for constant watering.

Flush only when necessary. Eliminating one flush per day—just one flush—could save your household 5,000 gallons per year.

Use the commercial car wash instead of soaping it up with a bucket and hose. Per wash? 100 gallons saved.

Eat less meat. Replacing one four-ounce serving of beef per week with veggies or soy can cut that virtual water we talked about by 20,000 gallons annually. If you're not into going completely veg, how about getting a single patty on that bun instead of two?

If you were to implement all of these suggestions—and they're not really difficult habits to manage—you'd be cutting your annual water consumption by 70,000 to 100,000 gallons or more per year.

This is where consciousness must take hold. This is where we matter. When a simple thing like turning off the faucet when you brush your teeth makes a big difference; or when choosing a salad instead of a sausage helps virtual water supplies, well, the choices don't seem so difficult.

No one is in our bathrooms, or spying on us at the supermarket. We have to choose the better option for the world out of something that comes from inside. Check it, there's a voice inside each of our heads that may not speak but gives pause. Within that moment, no matter how fleeting, we are given abandon. We can abandon bad habits, rote behavior, and make an informed decision.

So if I were to inform you that 25 percent of the water pumped into your house is literally flushed down the toilet, would you forego just one flush of pee? If you knew that the five gallons of water used for that one flush is as much as someone in a developing country uses for everything in their life for a whole day—drinking, washing, bathing—would you? Maybe. But even if you don't forego that one flush, the information may give you pause. You may decide to conserve water in some other way. This is the point.

HERE IN DULUTH it begins to snow. The moisture in the air is palatable. The gray sky's hues alert the world that something is about to change; precipitation will come.

Lake Superior will get to replenish its water. Levels will rise.

On my way to the airport, I speak with the driver of my shuttle bus. The weather news station is on. People pay close attention to the cold, wind, and snow. That famous "lake effect" on weather always has them bracing for the worst. The driver says it's been unseasonably warm, unseasonably dry. He's worried next summer the water problem could get worse, and that will affect him in all sorts of ways—personally, and professionally if the local economy slows because of water scarcity.

"Then again," he says, as I slide open the door and step out into the sleet that begins to come down, "it could get better." I smile and slide the door shut.

I'm heading home, these travels behind me, to figure out from that familiar perspective our effects on the miraculous web of the world. The consciousness around all that we do in our everyday lives awaits.

TEN

Where to Spread Your Wings

Home: Santa Monica, California

I open my eyes and along with 49,325,749 other people, experience the first sunlight of day. I am at home in California on Pacific Standard Time.

More than a quarter of a billion people in the United States have already experienced that same light. We can all now, if we so choose, share in the glory of the sun. We don't need as much man-made assistance. There is less need for lights and less need for heat; our energy costs can be lowered.

In theory, anyway.

We spend a million dollars per minute on energy in the United States. The sun can reduce that cost, but we don't use those rays to their full potential. Without even taking into account solar power, much of the sun's natural light is wasted. Daylight saving time allows us to capitalize on the earlier sunrises and later sunsets that occur during the summer months. By "springing forward" one hour, people in this country begin their days earlier; they can take advantage of the cool of the morning and get some work done before having to switch on their air conditioners. During summer evenings, the sun stays up longer; people spend more time outdoors and less time indoors with lights on; the availability of natural light later in the evening delays the need for artificial light. According to some estimates, this single additional hour of summer daylight allows us to trim 1 percent per day from the nation's summer electricity bill. Given that we spend about a trillion dollars per year on energy, we're saving almost $30 million a day simply by turning ahead our clocks. We don't extend that policy year-round. We should.

The idea behind daylight saving time was originally about saving money. You could even say it was about saving time, which is an intriguing proposition.

Benjamin Franklin is credited with the concept of daylight saving time, which he came up with while he was in Paris. He wrote an article called "An Economical Project," which posited that Parisians could save a great deal of money on wax candles if they would just get up earlier and work more during daylight hours. The idea became linked to coal and electricity during World War I when European countries adopted the scheme to conserve electric-power fuel. Today, we follow the spring-forward theory for some part of the year, but we fail to take full advantage of the benefits.

Extending daylight saving time year-round is a controversial proposal. Critics of this plan say it merely shifts more electrical consumption to mornings. In addition, it may even cause other types of energy usage—such as air conditioning in the afternoons because the sun is out longer and heating in the mornings when it's dark—to spike. Some also say more gas will be used because people will have more time on their hands . . . to drive around and shop. This is a peculiar argument. We can do our parts by not being piggish with heating and cooling and simply letting the sun do its job.

At home we waste about a quarter of our electric bills to keep the lights on (and the heat up) when the sun could do the job just fine. At work, it's much the same: 30 percent of energy is used to illuminate offices when there is usually sufficient natural light to turn many of the lights off.

Why do we choose artificial light over natural light? It's mostly out of habit. We just don't think about opening the shades and letting the sunshine in when we can flick a switch. Hell, we need remote controls to turn on our televisions, which are just a few feet away.

Our homes are where we spend the most time and where we use the most natural resources. Home is a sacred spot for many of us; a private place holding our possessions and material representations of who we are—our likes and dislikes.

We may believe that our homes are unique unto us—modern, traditional, historic—but where we hang our hats is pretty much the same all around the world. We have bedrooms and bathrooms and kitchens and places where we watch television that we call living rooms or family rooms.

The average US home today is more than 2,400 square feet. It uses about 100 million British thermal units (Btu) of energy (natural gas and electricity) at a cost of almost $1,500 per year. It consumes anywhere between 80,000 and 250,000 gallons of water and discards more than three tons of trash annually.

Our resource use and wasteful habits at home are where we can begin to make a big difference for this planet—and the effects will be felt all around the world as we've seen: from the palm oil in the cupboard to the Borneo rainforest, from the computer on your desk to mainland China, from the car idling in your driveway to the Arctic shores of Alaska, from the paper in your drawer to the Amazon jungle, from the water flowing from your tap to the Great Lakes, from the trash in your wastebasket to landfills like Fresh Kills, from your garbage disposal to the Pacific Ocean, and from your cell phone to India, you can trace the footprints of your activities all over the globe.

There is nothing on in my house now. No lights. No heat. No air conditioning. No appliances. No faucets. This is the bare elegance of life: uninterrupted warmth and comfort and the feeling of safety that only one's own home can somehow provide. There, in bed, I am.

But it's time to get ready for the day.

My water use, from my experiences in the last chapter, is as minimal as can be, yet I still manage in the bathroom to drain more than 10 gallons in my estimation: a single flush, a quick shower, and brushing my teeth will do that. Five gallons, the magic number, tallies fast: five gallons for every two minutes in the shower, five gallons for every flush, five gallons over the course of the day brushing your teeth if you keep the water flowing.

According to my water department, the average family of four living in my district uses about 163,000 gallons of water—about 2,608,000 glassfuls—a year for all of its water uses in and around the house.

The way to save water is simple: use less. Yet the force of habit for many people takes over. Is it necessary to keep the water running while you brush your teeth, or shave? Or to keep it going while you scrub dishes?

The water I'm using comes through pipes that snake their way into my basement. From there, they pop out onto the street and hook up with the main waterline. That waterline is managed by my city's water agency, which has even bigger pipelines leading to reservoirs. Those reservoirs have been filled by aqueducts stretching for miles, hundreds of miles that connect to rivers, streams, and lakes in Wyoming, Colorado, Utah, Arizona, Nevada, New Mexico, and in the rainier part of the state, the north coast and upper Sacramento Valley. It's the runoff from these sources that flows into the aqueducts.

Runoff occurs from snowmelt or rainfall. So the water I wash with could just have easily been snow that I skied on in the Sierra Nevada last winter.

The Metropolitan Water District explains that

raw or untreated water goes through several steps before it is ready to make its way through the pipes and into a home or business. First, the water flows through screens, which filter out leaves, sticks, fish, and other large debris. This can be done at a treatment plant or before it gets there. Once the water is filtered (screened), it gets a shot of chlorine to disinfect it and to help control taste and odor. Coagulants are added to the water to cause very fine particles to clump together into larger particles. *Flocculation* is a high energy mixing of the water that—just like using a blender to make a fruit smoothie—combines the particles. The water is then slowed down so the large particles formed in flocculation process settle out. The filters provide for removal of remaining particles and impurities from the water. The water then gets another dose of chlorine, to kill any leftover disease causing organisms and to give it a little extra protection for its next journey—to a reservoir.

Plumbing lines carry it from there to my home.

From my home, when water goes down the toilet or the drain, it ends up in the sewage system that captures wastewater and moves it more than eleven miles to a wastewater treatment facility. There, wastewater gets chemically treated and processed into fertilizer (otherwise known as sewage sludge), biogas (for use as energy), or pumped out to the ocean. I can't help, of course, but think of the *Alguita* that is still out at sea collecting particle samples from the Eastern Garbage Patch. Even though my treatment facility assures me and the rest of the public that "treatment capacity was expanded to prevent virtually all minute particles suspended in effluent from being discharged to the ocean environment" and that there have been "vast improvements in biological integrity of the bottom-dwelling marine community," it still makes me feel uncomfortable. I know that when I flush, or when my sink or tub drains, the water likely ends up floating out into the bay and from there . . .

There's my impact on the planet—maybe not too different from yours—within just ten minutes or so of waking up.

I'm conscious too of what I wear. Aren't we all? My choice this morning is an organic cotton T-shirt from the Gap, organic cotton jeans made by Levi's and my favorite Worn Again shoes.

I had to order the Worn Again shoes from London several years ago because they weren't yet available here in the United States; now they are. The shoes are made from scrap leather, ex-military jackets, reused buttons, and old T-shirts and jeans for the lining, as well as recycled rubber for the soles. Even the shoe box they come in is meant to be reused. On it there are five ideas:

1) Store your undies in it;

2) Use it as a durable yet elegant gift box;

3) Bury your dead pet hamster in it;

4) Create a window box and Grow Mustard, Cress and Alfalfa Sprouts;

5) Get snap happy and make a pinhole camera with it.

Wry British humor, for sure.

My T-shirt, while organic, makes me less comfortable. It is made in Sri Lanka, which is good for the Sri Lankan economy, but still, it has traveled a long, long way to land on my back. That means a lot of transportation energy was used in getting it to me. I could have chosen a regular cotton T-shirt made closer to home with lots of toxic agrochemicals, bleaches, and dyes. However, the Gap, I know, belongs to the Better Cotton Initiative (BCI), aimed at promoting more sustainable cotton cultivation practices worldwide. So the trade-off seems reasonable enough.

My Levi's "Eco" jeans (you can spot them because they have an embroidered lowercase "e" inside the front pocket, or at the bottom of the right leg of each jean), on the other hand, please me the most. While denim is notoriously bad for the environment because of the potency of its dyes and the energy and acidity used in its processing, these jeans are made from 100 percent organic fabric. They use recycled buttons, rivets, and zippers, as well as natural indigo for dye. They're made in the U.S.A.

Most of our clothing, believe it or not, is imported from Latin America. China sends us the second most, followed by other Asian countries. And those other Asian countries' exports are growing fast. My boxer shorts, for example, are made in Vietnam, where United States clothing imports grew from virtually nothing in 2000 to $3.2 billion as of 2006, making it the fifth-largest source of clothing imports to the United States behind China, Mexico, Indonesia, and India. My socks are from Honduras, where the sock business has boomed—doubling US exports to 28 million pairs a year—because of free-trade agreements. (However, President Bush recently signed legislation that puts Honduran sock imports in jeopardy.)

Look at the clothes in your own closet. I bet the majority of them come from the countries listed above. Only 10 percent of the apparel purchased in the United States is made here.

Despite my attempts to buy sustainably made clothing, it's rather difficult to do this and keep a wardrobe that doesn't look like it's suited for Woodstock. Normal-looking button-down shirts, suits, or dress slacks are hard to find in the sustainable clothing market. Remember, I said "normal."

The most environmentally-friendly clothing is clothing that you already own, or clothing that someone else has already owned. Thrift stores carry a huge inventory and variety of clothes. Still, they often come up short on proper sizing and selection. And who wants to always dress from some past decade of style?

Apparel is a large sector of the economy. We spend $345 billion a year on more than twenty billion pieces of clothing. That puts sixty-seven new garments into the average American's dresser and closet every year. The sustainable clothing industry is a tiny fraction of the retail business. Until demand grows for more organic or more sustainable textiles, manufacturers will offer limited fashion choices.

The best types of clothing—besides vintage, as I've mentioned—for the Earth are those that are grown organically (e.g., cotton, linen, hemp, wool); made from recycled fibers or materials; dye-free or use natural dyes; made in the United States or another first-world country that has strict environmental standards.

As far as what to avoid if you're trying to make more eco-friendly clothing choices, it's best to stay away from polyester and nylon, which are energy-intensive, polluting, and made from petroleum-based chemicals; nonorganic wool, as most wool is dipped in pesticides (to kill any insects that might be living in it); rayon, which is a wood-based fiber that requires an enormous amount of water and chemicals; furs of all types; and any garments that require dry cleaning.

I can—we can—make a simple difference by investigating where our clothes come from and whether they are made sustainably. It isn't easy figuring it out. Billions of dollars are spent by the fashion industry on advertising. Billions more are spent manufacturing and marketing their clothing. Millions of dollars are also spent by magazines and media outlets writing about style. A little more information about the origin and material of our apparel with which we can make more informed choices seems reasonable.

Fully dressed now in my kitchen, I open a box of cereal. But the rip-top plastic that is supposed to unveil the zip-lock top to my granola won't rip—well, with my hands anyway; I have to use my teeth. Standing there with that sheath of plastic in my mouth, I unscrew the plastic top to my soy milk. Then I have to pull the stopper out. Of course, the milk splatters and the cereal spills.

Each of us on average purchases and discards about two hundred pounds of plastic per year, and about sixty pounds of it is packaging that we just throw away. Considering that it takes a lot of manufacturing energy to make all that packaging, we're talking about a grand waste of resources, and I haven't even begun to think about the terrific trash trail that ensues.

We waste about ten cents of every dollar we spend on packaging, and we Americans spend four times more per person on stuff than any other country. Our packaging costs clearly add up to a lot of money. Beyond that, the manufacturing industry sucks up one-third of the energy and 13 percent of the water supply in the

United States, never mind all the waste—7.6 billion tons of it—that's produced before products even reach our hands. Of course, the products are then passed on to us in packaging of varying sorts and sizes, which is eventually tossed and recategorized as "household waste."

When products do hit our homes, our first task is usually to extract whatever it is that we actually wanted to buy—and toss away the bags and the packaging. Is it any wonder then why 1 trillion plastic bags are made every year, and that packaging comprises the most waste at landfills?

This leaves us as consumers with a responsibility. We need to be mindful of the things we buy (buying bigger sizes and in bulk saves the most packaging) and we need to recycle. But we also need to start pressuring manufacturers to cut down on packaging and to stop forcing waste upon us. That's right, complain.

Wal-Mart complained to its suppliers—I remember from investigating manufacturing issues in China—and it saved billions. My cereal box doesn't say where it's from, nor does my soy milk carton. But the cereal is distributed from Texas and the milk comes from Colorado. The coffee I've made is organic fair trade and comes from Ethiopia. That's pretty far away, but it supports farmers in that region and encourages responsible business practices.

Most food travels 1,500 to 2,500 miles before it makes its way into our bowls or onto our plates. In most cases it's better to buy locally grown produce and goods to save all the energy associated with transportation, cooling, and storage.

My milk carton can be recycled after removing the plastic stopper. So can my cereal box. When both are empty, I can put them in the recycle bin.

According to an article I read in the *New York Times* featuring an interview with the CEO of a large US cardboard supplier, most used cardboard fiber (such as cereal boxes) gets sent to

China where it's reprocessed and used to package goods that are sold . . . back to us.

A circuitous fate for my trash, as I look at it in the bin. I can almost picture the big message scrolled across the cartons: Less Packaging = Less Waste = Less Global Warming. I close the lid and decide to follow its fate further; I follow my trash when it's picked up in front of my house.

It's Monday. At 7:51 a.m. a trash truck pulls up in front of my house—as it does every week—and extends a large hydraulic fork around the blue recycle bin I've wheeled out onto the street. The fork clamps the bin and in a herky-jerky motion lifts it high and angles it over so the trash comes flowing out into the container bed. A second, two, three and then the claws place the empty bin back in the spot where it was so brusquely grabbed.

This morning I catch the driver's attention and tell him that I plan to follow him to the municipal recycling center. "No problem," he says, not even questioning this odd mission of mine. "I do a loop, until it's full. Maybe take a couple of hours," he explains.

"No problem," I say back, introducing myself and finding out that his name is Dwight and that he has worked for the city for seventeen years.

It's a crisp and clear day. The conditions are perfect for observing the subtleties of trash hauling.

I back down my driveway and catch up to Dwight who is now about six houses away. I follow him for a while right on his tail. Then I stay back, waiting for his move and gauging the traffic behind me. I feel like a private detective tailing a subject. I wait for cars to pass. I pull over in front of him, let him pass, park on the side of the road, hang back. This is mostly for my own entertainment because following a trash truck around first thing in the morning isn't my idea of a good time. Besides, I've had only one cup of coffee.

Dwight gets to a main road and takes a right. He goes up two blocks and takes another right where he begins to collect in the opposite direction of my street. He is picking up about one bin a minute. It's efficient work. The bins too are placed neatly and uniformly in a row; every house has dutifully put their trash bins out on the curb.

DO NOT FOLLOW
Frequent Stops
CAUTION
Wide Right Turns

I somnolently read the signs on the back of Dwight's truck. His hazards flash. He never goes above fifteen miles per hour. Occasionally he will jump out of the truck and fix a barrel or pick up trash that has fallen to the ground. And so it goes.

All around me the day's activities begin. A stressed school mom in a station wagon hurries two kids into the backseat of the car and races down the driveway in reverse. The kids barely have time to shut the door before she barrels out, nearly crashing into a Mercedes coming down the street. A father walks a son to school. The boy waves to Dwight. People wrangle their dogs, pick up after them, and gleefully talk or give them commands.

Two sheets of yellow paper escape a bin Dwight has lifted and get caught in the wind. They dance like kites for almost half a block.

Some people are anxious to get their bins off the street, and as soon as Dwight has emptied a barrel and places it back down, they appear, grab it, and roll it back up their driveway. A Muslim woman, donning a white headscarf, is late putting out her trash and she yells for Dwight to stop. He backs up and appeases her.

There are fewer homes on these streets and we travel faster. He travels down a street, takes a right at the end of a block, onto a boulevard, travels up that two blocks, skips a street, and then comes back down another boulevard, and then down one block to the street he skipped. Like this, in squares, up and down for two hours we go. Finally, Dwight stops and walks around the truck and approaches my car. He tells me that the truck is full and that he has to go dump it. He then explains how I am to follow him and what to do to get inside the gate of the recycling center.

We travel a mile away from my house to a weigh station where Dwight finds out that he has picked up 43,630 pounds of trash. He says he usually fills between two to two and a half loads a day. Then he travels another half-mile to the recycling center. This is the private recycling center where Adam Holt works and where I had visited on a previous occasion to get a formal tour. I hadn't known that my city contracted out its recycling business to this place. But now I know where my trash and recycling goes: "All over the world," as Adam explained. "Here" today, "over there" tomorrow.

Recycling in my district alone saves millions of tons of waste, not to mention dollars and energy. My city actually breaks this down by "environmental footprint"; and acres of land have been saved. It's a far cry from the slums of Dharavi and the recycling programs they have there. I now very much know the importance of recycling everything that I can, and buying with recycling or reusing in mind. Santa Monica is very "green" conscious and has a recycling rate more than double the national average.

Back at home I turn on my computer and begin my day of writing and researching. The black screen brightens and the machine powers on.

I correspond with the Middle East, Asia, Europe, India, and even read the *Alguita*'s "ship to shore" blog. The Internet, of course, allows me these opportunities. Yet more precisely the electric power coming out of my plug makes this possible.

Electricity works much like the Internet, in fact, with main lines servicing smaller sites that in turn service smaller applications that eventually winnow to a single use.

The electric outlet in my wall is wired to the meter "box" on the back of my house. From there, wires go underground and overhead to a transformer on a pole in back of the next house over. The transformer looks like a trash can bolted to a telephone pole twenty feet high. Wiring from there works its way to my local substation. My electricity provider services almost a thousand substations of various voltage levels to meet various needs. (Residential power use is 30 percent of the total energy allocated; industrial use is 40 percent; and commercial use is 30 percent, the electric company informs me.) The substations operate on a "power grid," so if one station's capacity to service is going over its limit, another can assist. They switch back and forth. The system is designed to create maximum efficiencies and avoid blackouts during peak-load hours.

The transmission lines from the substations to the power plants vary by resource. Two percent of my energy comes from biomass and waste; 10 percent comes from geothermal and small hydroelectric resources; 1 percent comes from solar power; 3 percent comes from wind; 8 percent comes from coal; 5 percent from large hydroelectric resources; 50 percent comes from natural gas; and 21 percent comes from nuclear power.

The power plants themselves are located as far away as Oregon and Utah. Power is also transmitted from the Hoover Dam in Nevada, and the Palo Verde nuclear power plant in Arizona.

The web of resources we use and the places from which they come are a miracle of human innovation and know-how. We've come a long way from fire. I am using energy that on average was created three hundred miles away.

As the day slogs on I am forced to use more electricity: the microwave (which is actually five times more energy-efficient than a traditional electric stove and three times more efficient than a gas oven). Lights too have to be flicked on so I can see. (I plug them in before I turn them on; any lamp or appliance that is plugged in even when turned off uses energy—all together, this "phantom power" can comprise as much as 10 percent of an electric bill.) The bulbs, of course, are compact fluorescent, which are 66 percent more energy-efficient and last ten times longer than traditional lightbulbs.

Daylight is fading, and the words I write—here—along with it. We are almost fully caught up now connecting the dots and linking together what we do and what that does to the planet.

I've explained the butterfly effect of water, waste, energy, and consumption. But I began this book by speaking about morality and the importance of knowledge and how that connects to appreciation. We need all three of these virtues in order to begin the process of change.

At the fulcrum of this book is an essay written in 1873 by Walter Pater, an Oxford University critic and poet. It basically says that we have an obligation—if to none other than ourselves—to be passionate, to be conscientious. It is our duty to experience, not merely observe, to reason and not take this world for granted.

> Not the fruit of experience, but experience itself, is the end. A counted number of pulses only is given to us of a variegated dramatic life. How may we see in them all that is to be seen by the finest senses? How shall we pass most swiftly from point to point, and be present always at the focus where the greatest number of vital forces unite in their purest energy? To burn always with this hard, gem-like flame, to maintain this ecstasy, is success in life. In a

sense it might even be said that our failure is to form habits: for, after all, habit is relative to a stereotyped world, and meantime it is only the roughness of the eye that makes any two persons, things, situations, seem alike. While all melts under our feet, we may well catch at any exquisite passion, or any contribution to knowledge that seems by a lifted horizon to set the spirit free for a moment, or any stirring of the senses, strange dyes, strange colors, and curious odors, or work of the artist's hands, or the face of one's friend. Not to discriminate every moment some passionate attitude in those about us, and in the very brilliancy of their gifts some tragic dividing of forces on their ways, is, on this short day of frost and sun, to sleep before evening . . .

We should not, as Pater says, "sleep before evening." We should not live in ignorance, but in a heightened state of awareness and understanding. It's about compassion for life. It's about appreciation for the world. By now you must understand that you are here, everywhere, and that your effect is felt.

I reach over and turn off the light. It is evening, time to sleep. The world goes dark. It is now all memory, and what more we make of it tomorrow.

Afterword

A sense of immunity has ferreted through our lives to the extent that many of us believe that nature exists far out in the woods. We've lost touch with the fact that the things we eat and drink, the homes we build, and the luxuries we enjoy are all natural. We all live in nature, and we all live off what nature provides. We may fashion things as man-made, but we can't take credit for the ingredients. Connecting resources back to their habitats is part of what we've done in this book, and hopefully that has drawn us closer to understanding the exploitations we extend.

The power we hold over the planet is awesome. We have changed the Earth's natural course of development and affected its health by what we do and how we live our lives. We can just as easily change its course again—for the better.

Embracing knowledge and creating awareness can reshape our lives and make the future more certain.

The world is not, I repeat not, going to end anytime soon. We will figure out ways to fix the environmental problems facing us, even if we have to go to extremes, such as geo-engineering a parasol over the planet, building giant carbon filters, or continuously spraying ocean water into the air to cool the Earth's temperature. Believe it or not, these are all actual solutions being contemplated.

It shouldn't have to come to any of that. We can exercise compassion—our wonderful ability to care—and change our actions, which in turn will change consequences.

I'm neither a fearmonger nor a Pollyanna pusher. In truth, the most powerful and eye-opening thing I discovered in my travels is how much compassion we have for each other and the planet—no matter the geographic location or nationality of the people. Bundle us all up and we can force change of unimaginable magnitude.

The way in which I personally live now has greatly changed. Almost every action I take is with heightened consideration. Before this book I was, of course, green conscious. I could blather on about all the environmental data points. But now I see *people* in my actions. I associate differently. I *feel* differently about what I do and what that does to the planet. This is a huge shift in my awareness.

What has seeped through my experience and my investigation of the world's most troubled environments are the connections I made with the people and places most affected by climate change. The consequences of environmental crises are no longer surface or anecdotal—they are people's lives.

This is what sits deeply with me now. These are the reminders that not only keep me mindful but incite me to do even more to save our natural resources and think of new ways we can just plain do things smarter. So many problems can be fixed by first being made aware that they exist, and then finding out what we are doing to cause them. Solutions abound.

I can now visualize my green effect, my carbon footprint, and my strain on natural resources. I understand my place in this ecological conundrum and am empowered by what I know. It is my hope that after reading these pages, you do too.

We now know how much we all matter. So what are we going to do with this understanding?

Acknowledgments

This book could not have been done without the assistance of Dr. Colleen Howell. It would not have been born without Gideon Weil. And certainly Susan Raihofer deserves so much more credit than this one line gives. I am also lucky to have Margaret Riley in my life. More thanks and appreciation go to Suzanne Biegel, Gary Brooks, and the team at ProTravel, Jeffrey Dash, Joy Fehily, Sara Vicendese, Lauren Auslander, Matt Hurwitz, Brendon Sher, Matt Rees, Debi Goenka, Tony Weyiouanna Sr., Mathew Maavak, Katy Wong, Derrick Boelter, Hila and Jeremy Wenokur, Jay Lawrence Goldman, Jill Siegel, the staff at Ariau Amazon Towers, Reality Tours & Travel, Ambassador Passport and Visa, Robert Lawrence, the crew of the *Alguita*, Adam Holt, the city of Santa Monica, and Craig and Beth Malloy. Others who helped this book come to be include Mark Tauber, Claudia Boutote, Michael Maudlin, Jan Weed, Emily Grandstaff, Terri Leonard, Alison Petersen, and Laina Adler. For my tools and gear: Dell, The

North Face, Panasonic, BIC, Moleskine, Timex, Blackberry, and AT&T. For putting me up: Shishmaref Emergency Station, the Dan Hotels, Hotel Marine Plaza, the Hilton Hotels, Tang Yao Hotel, Shanxi Grand Hotel, Holiday Inn, and the Chelsea Hotel. Last but not least, I'd like to thank my family and friends for giving me something to come home to.

Resources

We make no claims for nor do we warrant credit for the research provided by other sources. They deserve it. We believe we've captured them all, but forgive us if something has fallen through the cracks. We're making room on *You Are Here*'s Web site (www.readyouarehere.com) for updates, posts, and comments. We encourage dialogue; that is what this book is designed to create. These issues are all of ours.

CHAPTER 1: LOSING OUR PAST

If everyone on Earth lived like those in America, we'd need five planets to support us.
*http://www.loe.org/shows/segments
    htm?programID=07-P13-00045&segmentID=2*

Eight hundred and fifty million people worldwide live with chronic hunger.
http://www.weforum.org/en/initiatives/hunger/index.htm

More than 1 billion people worldwide lack access to safe drinking water.
http://www.usaid.gov/our_work/global_health/eh/news/wwd2008.html

One second of the sun's energy could power the US for 9 million years.
http://www.swpc.noaa.gov/primer/primer.html

Food scraps are the third most common type of waste in the US.
http://www.epa.gov/epaoswer/non-hw/muncpl/pubs/msw06.pdf

Concentration of CO_2 2,500 years ago was approximately 280 ppm. Current concentration is roughly 384 ppm.
*http://cdiac.ornl.gov/ftp/trends/co2/vostok.icecore.co2.
    http://www.earthpolicy.org/Indicators/CO2/2008.htm*

Greenhouse gases and global warming explained
*http://lwf.ncdc.noaa.gov/oa/climate/globalwarming.html
    http://earthguide.ucsd.edu/globalchange/global_warming/03.html*

Climatic impacts of global warming
http://www.unep.org/themes/climatechange/PDF/ipcc_wgii_guide-E.pdf

Acid rain facts
http://www.epa.gov/acidrain/index.html

Acid rain's effects on ancient structures
http://www.geotimes.org/july07/article.html?id=geophen.html

Salt weathering and climate change
http://www.buildingconservation.com/articles/atmospheric/atmospheric.htm

World Monument Fund 2008 Watch List press release
http://wmf.org/pdf/Watch_2008_release.pdf

Climate change already causing destruction to historic sites
http://whc.unesco.org/uploads/news/documents/news-262-1.doc

South Africa's 117,000-year-old fossilized footprints
http://www.ecoafrica.com/african/travel/WestCoast.html

Twelve thousand Scottish sites threatened by erosion and sea level.
*http://www.unep.org/Documents.Multilingual/Default.
    asp?DocumentID=485&ArticleID=5412&l=en*

More than 80 percent of China's 33 World Heritage sites damaged by air pollution and acid rain
http://www.usembassy-china.org.cn/sandt/estnews-July-August2005.pdf

Peru's pre-Inca treasures threatened by glacial melt
http://www.unep.org/Documents.Multilingual/Default. asp?DocumentID=485&ArticleID=5412&l=en

Total savings from turning thermostat up one degree in the summer is calculated by estimating that average US household uses about 2,800 kWh per year on air conditioning. A one-degree increase could result in an energy reduction of as much as 5 percent, or 140 kWh per year per household. Across the estimated 60 million households in the US who have air conditioners, this is 8.4 billion kWh saved per year.
http://www.eia.doe.gov/emeu/recs/recs2001/enduse2001/enduse2001. html#table2
http://www.epa.gov/climatechange/emissions/ind_calculator.html

Total savings from turning thermostat down one degree in winter calculated by estimating that average US household with gas heating uses 53 thousand cubic feet of natural gas per year on heating. A one-degree decrease could result in an energy reduction of as much as 5 percent, or 2,650 cubic feet per year. Across the estimated 61 million households in the US who use gas heating, this is 161.6 billion cubic feet per year.
http://www.eia.doe.gov/emeu/recs/byfuels/2001/byfuel_ng.pdf
http://www.epa.gov/climatechange/emissions/ind_calculator.html

Total reduction in CO_2 from thermostat adjustments calculated using emissions rate of 1.37 pounds of CO_2 per kWh and 120.61 pounds CO_2 per thousand cubic foot of natural gas
http://www.epa.gov/climatechange/emissions/ind_calculator.html

The average coal-fired power plant emits about 5 million short tons of CO_2 per year.
http://www.epa.gov/grnpower/pubs/calcmeth.htm#coalplant

Sulfur-dioxide emissions estimated at 6 pounds per 1000 kWh, on average
http://www.epa.gov/appdstar/pdf/brochure.pdf

Teleconnections
http://www.cpc.ncep.noaa.gov/data/teledoc/teleintro.shtml

Definition of "climate" and "weather"
Merriam Webster's Collegiate Dictionary, tenth edition

Carbon second-most common element in human body
http://www.newworldencyclopedia.org/entry/Human_body

CHAPTER 2: OUR FUTURE

Greater Mumbai Urban Agglomeration population (2007 est.)
http://www.prb.org/Articles/2007/delhi.aspx?p=1

Seventy-one percent of India's population is rural.
http://www.ruralpovertyportal.org/english/regions/asia/ind/statistics.htm

More urban dwellers than rural dwellers for the first time in history
http://news.ncsu.edu/releases/2007/may/104.html

Global population residing in megacities
http://www.ciesin.columbia.edu/documents/vulofglob_contactshtml.pdf

More than half of the US population lives within 50 miles of the shore, and coastal cities have grown by nearly 30 percent over the past 25 years.
http://www.magazine.noaa.gov/stories/mag167.htm

Seventy-five percent of Mumbai sewage discharged without being treated
http://www.ciesin.columbia.edu/documents/vulofglob_contactshtml.pdf

Squatter community density and latrines per capita
http://www.ciesin.columbia.edu/documents/vulofglob_contactshtml.pdf

Mumbai health problems due to pollution
http://www.newint.org/issue290/volcano.htm

Eight hundred and eighty-four million gallons piped in to Mumbai per day; 20 percent wasted
http://www.idswater.com/water/us/News/2284/pressrelease_content.html

Mumbai is predicted to experience an average annual temperature increase of 2.25 to 3.25 degrees Fahrenheit, and a sea level rise of 50 cm.
http://www.ciesin.columbia.edu/documents/vulofglob_contactshtml.pdf

2005 monsoon kills more than 1,000 in Mumbai
http://www.newscientist.com/article.ns?id=dn7768

Mumbai geography
http://www.indiaholidaystours.com/mumbai-holidays.htm

Flooding problems with monsoons and rail transport
http://www.ciesin.columbia.edu/documents/vulofglob_contactshtml.pdf

Ten percent of the global population lives in low-lying areas and 130 of the world's 180 coastal nations have built cities in these areas.
http://www.iied.org/mediaroom/releases/070328coastal.html

Ten countries with largest populations living in low-lying areas
http://www.sciencedaily.com/releases/2007/03/070328093605.htm

Conservation Action Trust, Mumbai
http://www.karmayog.com/ngos/cat.htm

Seventy thousand taxis and autorickshaws in Mumbai
http://www.rediff.com/money/2007/may/18cab.htm

History of Mumbai
http://www.encyclopedia.com/doc/1B1-357646.html

Fifty to 80 percent of electronics collected for recycling are shipped overseas.
http://www.ban.org/E-waste/technotrashfinalcomp.pdf

E-waste exports to India
http://www.atimes.com/atimes/South_Asia/HH03Df01.html

The US generates about 2.9 million tons of e-waste annually (2006 est.) and
about 330,000 tons are recycled. With 80 percent of recycled e-waste
being shipped overseas, this is roughly 264,000 tons that are exported.
http://www.epa.gov/epaoswer/non-hw/muncpl/pubs/06data.pdf
http://www.ban.org/E-waste/technotrashfinalcomp.pdf

The tonnage of e-waste we toss out with the garbage each year—about 2.57
million tons, or 4.6 billion pounds—is equivalent to about 685 million
laptops (assuming 7.5 pounds per average laptop).
http://pcworld.about.com/od/notebooks/Ditch-Your-Desktop-for-a-Lapto.htm

Fifty million tons of e-waste generated per year
http://www.unep.org/Documents.Multilingual/Default.
asp?DocumentID=485&ArticleID=5431&l=en

Ravi Agarwal quote in Basel Action Network's Toxic Trade News
http://www.ban.org/ban_news/2007/071114_global_waste_destination.html

Photos of LA County School District tags on e-waste
http://www.ban.org/photogallery/china_guiyu/pages/losangeleslabels_pic.html

BAN Web site quote
http://www.ban.org/main/about_BAN.html#whytoxic

Health impacts of electronic waste
http://www.etoxics.org/site/PageServer?pagename=svtc_lifecycle_analysis

Assuming an average density of 220 pounds per cubic yard of electronic
waste, 264,000 tons of e-waste would fill a volume of about 2.46 million
cubic yards. Divide this by 4,840 square yards per acre and the height of
the acre-wide pile is 508 yards (1,524 feet) high.
http://www.cambridgema.gov/TheWorks/departments/recycle/pdffiles/pla-
ninstructions.pdf

The Empire State Building is 1,454 feet to the top of the lightning rod.
http://www.newyorktransportation.com/info/empirefact2.html

Per-capita waste generation in the US from 1967 to 2006 has gone from
about 3 pounds to 4.6 pounds per day. About 32.5 percent is recycled.
http://www.epa.gov/epaoswer/non-hw/muncpl/pubs/msw06.pdf

Plastic bottle and aluminum can use in the US
   http://www.container-recycling.org/

Paper cup use in North America
   http://www.paper360.org/paper360/data/articlestandard//
      paper360/252007/436364/article.pdf

US recycling newspaper recycling and disposal rates
   http://www.epa.gov/epaoswer/non-hw/muncpl/pubs/06data.pdf

Total waste generated per year in the US
   http://www.epa.gov/epaoswer/non-hw/muncpl/pubs/msw06.pdf

Enough hazardous waste generated per year to fill the Superdome 1,500
times
   http://www.cleanair.org/Waste/wasteFacts.html

Dharavi slum
   http://ngm.nationalgeographic.com/ngm/0705/feature3/index.html

Dharavi produces $650 million per year of products.
   http://www.millenniumassessment.org/documents_sga/indian%20urban%2
      0sa%2030%20pager.pdf

Openly burning trash emits 10,000 times more pollution than is emitted
when trash is burned in a regulated incinerator.
   http://www.dioxinfacts.org/sources_trends/trash_burning.html

"… the noblest conception on earth is that of men's absolute equality."
   Ayn Rand. 1943. *The Fountainhead.* New York: The Bobbs-Merrill
   Company.

CHAPTER 3: WE ARE NOT ALONE

Borneo home to most species on Earth
   http://www.physorg.com/news4072.html

WWF summary of types and numbers of species in Borneo
   http://assets.panda.org/downloads/treasureislandatrisk.pdf

More species in Borneo than in all of North America
   http://www.ciesin.columbia.edu/docs/002-256a/002-256a.html

Malaysia and Indonesia are biggest tropical timber exporters in the world.
   http://www.itto.or.jp/live/Live_Server/377/E-AR06-Text.pdf

Indonesia now leading palm oil producer
   http://www.pecad.fas.usda.gov/highlights/2007/12/Indonesia_palmoil/

Malaysia and Indonesia supply the world with 34 million metric tons of
palm oil.
   http://www.pecad.fas.usda.gov/highlights/2007/12/Indonesia_palmoil/

At least half of palm plantations are associated with deforestation.
http://www.foe.co.uk/resource/reports/greasy_palms_summary.pdf

Palm is second most used edible oil after soy.
http://www.foe.co.uk/resource/reports/greasy_palms_summary.pdf

Lowland forests in the Indonesian region of Borneo to be completely cleared by 2010 and upland forests will disappear by 2020
*http://wbln0018.worldbank.org/eap/eap.nsf/Attachments/
Indonesia+Environmental+Report/$File/Indonesia+ENVNRM+Transit
ion-entire.pdf*
http://www.worldwildlife.org/wildplaces/borneo/marketforces.cfm

Greenpeace prevents shipment of palm oil.
http://www.greenpeace.org/usa/news/palm-oil-blockade

Roundtable on Sustainable Palm Oil
http://www.rspo.org/About_Sustainable_Palm_Oil.aspx

Deforestation is the second-biggest man-made contributor to climate change after burning fossil fuels, and accounts for about 20 percent of carbon dioxide emissions per year.
http://earthtrends.wri.org/updates/node/266

Carbon emissions statistics from Borneo deforestation were calculated with the following assumptions. First, no carbon data exists for Borneo by itself, because it's split between 3 countries. So these figures are based on aggregated data. According to the FAO as cited by *http://news.mongabay.com/2006/1105-indonesia.html* the destruction of 1.87 million hectares of Indonesian forests releases 94 to 281 (average = 187) million metric tons of carbon per year, or about 50 to 150 metric tons per hectare. Borneo's deforestation rate is 1.3 million hectares per year.
http://assets.panda.org/downloads/treasureislandatrisk.pdf.

So converting from carbon to carbon dioxide, on the high end, we can estimate that Borneo's deforestation may result in 715 million short tons of carbon dioxide per year.

The average vehicle releases 6 tons of CO_2 per year.
http://www.epa.gov/climatechange/emissions/ind_calculator.html

The Indonesian government estimates some 90 percent of logging in the country is against the law.
http://www.ens-newswire.com/ens/feb2004/2004-02-06-11.asp

Effluents from palm oil refineries damaging marine environment
http://www.aseanbiodiversity.info/Abstract/53005299.pdf

One hundred twenty million people dependent on fisheries from Coral Triangle
http://www.nature.org/pressroom/press/press3127.html?src=search

WWF: "If our oceans die, so do we"
*http://www.wwf.org.my/how_you_can_help/donate_main/save_our_
oceans_/index.cfm*

Multimedia Super Corridor (MSC) Malaysia
http://www.kiat.net/msc/

Orangutan is Asia's only great ape.
*http://www.panda.org/about_wwf/what_we_do/species/about_species/
species_factsheets/great_apes/index.cfm*

Orangutan means "man of the forest."
*http://www.panda.org/news_facts/publications/index.
cfm?uNewsID=62940*

Orangutans often killed for eating palm saplings
*http://redapes.org/palm-oil/activists-palm-oil-workers-killing-endangered-
orangutans/*

Orangutan population status
http://www.foe.co.uk/resource/reports/oil_for_ape_full.pdf
*http://www.asiasentinel.com/index.php?option=com_content&task=
view&id=381&Itemid=34*

Status and benefits of mangroves
http://www.fao.org/newsroom/en/news/2008/1000776/index.html

Market prices for timber in Sarawak
*http://www.ihb.de/madera/news/Malaysia_Monsoon_season_price_hike_
15847.html*

An old-growth tree can contain 1,000 or more cubic feet.
http://fax.libs.uga.edu/text/2tgbitxt.txt

An average tree containing one cord (128 cubic feet) of wood will produce
almost 90,000 sheets of paper.
http://www.ecology.com/feature-stories/paper-chase/index.html

Palm oil facts
http://www.americanpalmoil.com/faq.html#1

Health effects of eating palm oil
http://www.cspinet.org/new/pdf/palm_oil_final_5-27-05.pdf

Palm oil often labeled as "vegetable oil"
http://www.borneoproject.org/article.php?id=60

WWF Roundtable on Sustainable Palm Oil
*http://www.panda.org/about_wwf/what_we_do/forests/our_solutions/
responsible_forestry/forest_conversion_agriculture/roundtables_soy_
palmoil/roundtable_on_sustainable_palm_oil/index.cfm*

Species statistics
   http://www.panda.org/about_wwf/what_we_do/species/problems/index.cfm

E. O. Wilson. 1998. *Consilience: The Unity of Knowledge.* New York: Random House.

CHAPTER 4: WHAT ARE WE DOING?

Linfen, China, the dirtiest place on earth
   http://www.blacksmithinstitute.org/site10b.php

China is building a new coal power stations every four days.
   http://www.wilsoncenter.org/topics/docs/transboundary_feb2.pdf

China exports approaching $1 trillion
   http://www.intracen.org/appli1/TradeCom/TP_EP_CI.aspx?RP=156&YR=2006

China is world's leading carbon emitter
   *http://www.mnp.nl/en/dossiers/Climatechange/moreinfo/Chinanowno1in
      CO2emissionsUSAinsecondposition.html*

IPCC recommendations to reduce greenhouse gas concentrations
   http://www.ipcc.ch/pdf/assessment-report/ar4/wg3/ar4-wg3-spm.pdf

An average of 25 percent of the pollution in China is from producing goods
   for export.
   http://pubs.acs.org/subscribe/journals/esthag-w/2006/mar/policy/kc_china.html

Thirty-five percent of China's CO_2 emissions are from production of goods
   for export according to International Energy Agency's annual World
   Energy Outlook publication for 2007.
   http://www.lowtechmagazine.com/2008/01/the-worlds-fact.html#more

On some days, one-quarter of LA air pollution linked to coal plants in China
   *http://www.climate.noaa.gov/index.jsp?pg=/news/news_index.
      jsp&news=story_chinagprus.html*

US imports from China more than $320 billion in 2007
   http://www.census.gov/foreign-trade/balance/c5700.html#2008

Top commodities imported to the US from China
   *http://ia.ita.doc.gov/trcs/monitoring/china/imports/output/china4digit_
      yty1.html*

Wal-Mart imports $27 billion worth of goods from China each year.
   http://www.epi.org/content.cfm/ib235

Wal-Mart press release about packaging reduction initiative
   http://www.walmartstores.com/FactsNews/NewsRoom/5951.aspx

Thirty thousand Chinese factories supply Wal-Mart.
   http://www.edf.org/page.cfm?tagID=1458

China GDP growth and inflation figures
http://www.forbes.com/markets/feeds/afx/2008/04/08/afx4864410.html

Guangdong Province threatens lawsuit.
http://www.caltradereport.com/eWebPages/front-page-1194554120.html

1984 Los Angeles Summer Olympic Games ground-level ozone reduction
http://www.epa.gov/oar/epa450.txt

1996 Atlanta Summer Olympic Games ozone reduction
http://jama.ama-assn.org/cgi/reprint/285/7/897.pdf

Wired magazine on China's plan for a green Olympics
*http://www.wired.com/science/planetearth/magazine/15-08/ff
_pollution?currentPage=2*

Top 30 imported products from China based on total value
*http://ia.ita.doc.gov/trcs/monitoring/china/imports/output/china4digit_
yty1.html*

List of toy recalls
http://www.cpsc.gov/cpscpub/prerel/category/toy.html

Firms move to China to take advantage of cheap labor and lax environmental laws.
http://www.globalpolicy.org/socecon/envronmt/climate/2007/1012south.htm

Linfen River second-longest tributary to Yellow River
http://china.travelmixture.com/city/Linfen.htm

Yellow River one of world's most polluted rivers
*http://www.iwmi.cgiar.org/assessment/FILES/pdf/publications/Research
Reports/CARR3.pdf*

Linfen clinics reporting growing cases of respiratory diseases. Lead poisoning on the rise
http://www.blacksmithinstitute.org/site10b.php

Quote from Linfen resident to *The World* reporter
http://www.theworld.org/?q=node/4059

Linfen death rates
http://www.npr.org/templates/story/story.php?storyId=10221268

More than one in three Americans lives in an area with unhealthy air.
http://lungaction.org/reports/sota07exec_summ.html

Asthma statistics
http://www.cdc.gov/mmwr/preview/mmwrhtml/ss5608a1.htm

Lake Tai pollution, *New York Times*
*http://www.nytimes.com/2007/10/14/world/asia/14china.
html?_r=1&hp&oref=slogin*

US signs Kyoto Protocol, but never ratifies it.
http://maindb.unfccc.int/public/country.pl?country=US

Kyoto Protocol
http://unfccc.int/kyoto_protocol/items/2830.php

Key elements of the Bali roadmap
http://news.bbc.co.uk/2/hi/science/nature/7146132.stm

Health effects from coal pollution in China expected to total $39 billion by 2020
http://assets.panda.org/downloads/coming_clean.pdf

Air pollution hindering China's economic growth by 3 to 6 percent of GDP annually
http://www.wilsoncenter.org/topics/docs/transboundary_feb2.pdf

Study reports that improving air quality in China would very slightly increase the cost of goods
http://pubs.acs.org/subscribe/journals/esthag-w/2006/mar/policy/kc_china.html

Coal in Shanxi Province provides China with two-thirds of its energy.
http://www.citymayors.com/environment/world_pollution.
html#Anchor-Linfen-23240

Erin Cline Davis. "It's worse than dirty. Dirty air has toxic components; L.A.'s notorious air pollution is hardest on kids. The closer to a freeway they live, play or attend school, the more likely it is that their developing lungs' capacity will be reduced." December 10, 2007. *Los Angeles Times.*

Population of Linfen
http://www.chinadaily.com.cn/china/2007-05/16/content_873507.htm

Pollution in China highest during the winter
http://pubs.acs.org/subscribe/journals/esthag/40/i15/html/080106news1.html

"Rising sea levels to take toll on nation"
http://en.chinagate.com.cn/environment/2007-11/23/content_9282101.htm

"Provinces construct green energy plants"
http://www.chinadaily.com.cn/bizchina/2007-12/05/content_6299525.htm

"Waging a tough war against plastic bags"
http://www.chinadaily.com.cn/cndy/2007-11/24/content_6276166.htm

National Action Plan for Energy Efficiency
http://www.epa.gov/cleanrgy/documents/napee/napee_exsum.pdf

A ton of coal costs between about $50 and $100.
http://www.eia.doe.gov/cneaf/coal/page/coalnews/coalmar.html#weekly

"The average emission rates in the United States from coal-fired generation are: 2,249 lbs/MWh of carbon dioxide, 13 lbs/MWh of sulfur dioxide,

and 6 lbs/MWh of nitrogen oxides." One ton of coal supplies about 2.1 megawatt-hours of electricity.
http://www.epa.gov/cleanenergy/energy-and-you/affect/coal.html
http://www.epa.gov/appdstar/pdf/brochure.pdf

One ton of coal—2100 kWh—could light about 400 homes for a day. The average US home uses 1950 kWh per year for lighting.
http://www.energystar.gov/ia/partners/promotions/change_light/downloads/ CALFacts_and_Assumptions.pdf

John Holdren quote
http://discovermagazine.com/2006/dec/clean-coal-technology

US, China, and India will collectively be emitting 2.7 billion tons of carbon dioxide per year by 2012.
http://discovermagazine.com/2006/dec/clean-coal-technology

Canadian Clean Power Coalition describes how coal works.
http://www.canadiancleanpowercoalition.com/Customer/ccpc/ccpcwebsite. nsf/AllDoc/9D9A829BA5468A3A8725697B007D2DA2?Open Document

A single coal plant can generate close to 500,000 tons of fly ash.
http://www.greatriverenergy.com/about/coal_plants.html

US is often referred to as the Saudi Arabia of coal.
http://www.csmonitor.com/2006/0710/p02s01-usec.html

US coal exports
http://www.eia.doe.gov/cneaf/coal/quarterly/html/t7p01p1.html

Cost of clean coal plant vs. conventional plant
http://discovermagazine.com/2006/dec/clean-coal-technology

IGCC process and collection and burial of emissions
http://discovermagazine.com/2006/dec/clean-coal-technology

Intergovernmental Panel on Climate Change estimate of how much carbon dioxide can be stored in old coal mines. Enough storage for 320 years of coal-power carbon emissions.
http://discovermagazine.com/2006/dec/clean-coal-technology

Incineration for energy and comparison to coal
http://www.chinadaily.com.cn/hkedition/2007-10/29/content_6211957. htm

Chinese garbage-burning plant incinerates 1,040 tons of waste per day to generate 130 million kWh of energy.
http://www.ccchina.gov.cn/en/NewsInfo.asp?NewsId=9556

Estimated cost to repair world's energy infrastructure
http://discovermagazine.com/2006/dec/clean-coal-technology

Air movement and the Coriolis Effect
  http://www.windows.ucar.edu/tour/link=/earth/Atmosphere/transport.
  html&edu=high

CHAPTER 5: DISTANT CONSEQUENCES

Shishmaref erosion rate 10 feet per year
  http://www.csc.noaa.gov/cz/2007/Coastal_Zone_07_Proceedings/PDFs/
  Tuesday_Abstracts/3410.Gray.pdf

Shishmaref village site "lost to the sea" by 2020
  http://www.csc.noaa.gov/cz/2007/Coastal_Zone_07_Proceedings/PDFs/
  Tuesday_Abstracts/3410.Gray.pdf

Shishmaref income and poverty statistics
  http://www.commerce.state.ak.us/dca/commdb/CIS.
  cfm?Comm_Boro_Name=Shishmaref

Contribution of melting ice caps and glaciers to sea level rise
  http://www.nsf.gov/news/news_summ.jsp?cntn_id=109759

US Army Corps of Engineers assessment
  http://edocket.access.gpo.gov/2003/pdf/03-29010.pdf

Polar regions and temperature increases due to greenhouse gases
  http://www.weather.nps.navy.mil/~psguest/polarmet/clichange/index.html

Greenland's ice loss; Ayles ice shelf; Crack in ice pack from Russia to the
  North Pole
  McKenzie Funk, "Cold Rush: The Coming Fight for the Melting
  North," *Harper's,* Sept. 2007 (subscription required).

University of California, Irvine, study on dirty snow
  http://today.uci.edu/news/release_detail.asp?key=1621

Asian sources of soot and ash deposited on Alaska and the Arctic
  http://www.pubmedcentral.nih.gov/articlerender.fcgi?artid=327163

Average carbon dioxide emissions per adult per year
  http://www.epa.gov/climatechange/emissions/ind_calculator.html

Average air-travel carbon dioxide emissions per mile
  http://yosemite.epa.gov/oar/globalwarming.nsf/UniqueKeyLookup/
  LHOD5MJTA3/$File/2003-final-inventory_annex_e.pdf

Average vehicle travel carbon dioxide emissions per mile (assume about 20 mpg)
  http://www.climatecrisis.net/takeaction/carboncalculator/howitwas
  calculated.html

Average carbon dioxide emissions per kWh
  http://tonto.eia.doe.gov/ftproot/environment/e-supdoc-u.pdf

Average carbon dioxide emissions and fuel usage for natural gas and heating oil
http://www.epa.gov/climatechange/emissions/ind_calculator.html

Percentage of US energy generated from coal
http://www.epa.gov/cleanenergy/energy-resources/egrid/index.html

Roughly half of the carbon dioxide emitted into the atmosphere each year is absorbed by the Earth's natural processes. The other half contributes to warming.
http://www.eia.doe.gov/oiaf/1605/ggccebro/chapter1.html

Dioxins from incinerators found in Arctic regions
http://www.cec.org/files/PDF/POLLUTANTS/dioxexec_EN.pdf

Unequal sex ratios in Arctic communities from PCB poisoning
ftp://ftp.cordis.europa.eu/pub/sustdev/docs/environment/polar_symposium_all_abstracts_final_en.pdf

Hermaphrodite polar bears
http://www.ehponline.org/members/2003/5553/5553.html

Intercontinental transport of air pollution
http://esrl.noaa.gov/csd/AQRS/reports/itap.pdf

Pesticide travel via ocean currents
http://www.columbia.edu/~pjs2002/arctic/pages/pollution.html

Albedo
http://www.pbs.org/saf/1404/features/thermostat.htm

Arctic haze
http://esrl.noaa.gov/csd/AQRS/reports/itap.pdf

National Science Foundation sea-level-rise statistics
http://www.nsf.gov/news/news_summ.jsp?cntn_id=109759

A one-food sea-level rise typically causes a shoreline to retreat about 100 feet.
http://www.nsf.gov/news/news_summ.jsp?cntn_id=109759

One hundred million people live within three feet of sea level.
http://www.nsf.gov/news/news_summ.jsp?cntn_id=109759

More than 500 million people live in low-elevation coastal areas.
http://www.iied.org/mediaroom/releases/070328coastal.html

More than half the US population lives within 50 miles of a coast.
http://coastalmanagement.noaa.gov/partnership.html

CHAPTER 6: NATURE'S OXYGEN FACTORY

The Amazon accounts for about 17 percent of the world's forested area.
http://www.worldlandtrust.org/news/carbonfigures.pdf

Locked-up in the Amazon is 95 billion tons of carbon—equivalent to 11 years' worth of carbon from human activities.
http://news.mongabay.com/2007/0508-amazon.html

Deforestation comprises 20 percent of human-caused greenhouse gas emissions.
http://news.mongabay.com/2007/0510-red.html

Current deforestation rates in the Amazon (as an average of 2002–2007) equal 4.8 million acres per year, or 7505 square miles.
http://www.mongabay.com/brazil.html

Massachusetts covers 7,840 square miles.
http://quickfacts.census.gov/qfd/states/25000.html

Amazon deforestation (2000-2007) equals 59,081 square miles (37.8 million acres).
http://www.mongabay.com/brazil.html

Georgia covers 57,906 square miles.
http://quickfacts.census.gov/qfd/states/13000.html

Declining air quality over past century
Van Aardenne, J. A., F. J. Dentener, J. G. J. Olivier, C. G. M. Klein Goldewijk, and J. Lelieveld (2001) A 1x1 degree resolution dataset of historical anthropogenic trace gas emissions for the period 1890–1990. *Global Biogeochemical Cycles,* 15(4), 909–928.

NO_x health effects
http://www.epa.gov/oar/urbanair/nox/noxfldr.pdf

Four million six hundred thousand people die from air pollution each year—more than the number who die from automobile accidents.
http://www.sciencedaily.com/articles/a/air_pollution.htm

In terms of particulate matter, Venezuela has the best air quality and China has the worst.
http://siteresources.worldbank.org/DATASTATISTICS/Resources/table3_13.pdf

Five major air pollutants regulated by the Clean Air Act
http://www.epa.gov/air/urbanair/6poll.html

Best air quality for US cities in Cheyenne, WY
http://lungaction.org/reports/sota07_cities.html#5a

Worst air quality for US metropolitan areas in Los Angeles–Long Beach–Riverside, CA
http://lungaction.org/reports/sota07_cities.html#2

Los Angeles worst traffic in nation
http://tti.tamu.edu/documents/mobility_report_2007.pdf

Whittier, CA, in Los Angeles County is home to the nation's largest landfill.
http://www.laalmanac.com/environment/ev04.htm

About 230,000 square miles of Amazon rainforest have been lost to deforestation since 1970.
  http://www.mongabay.com/brazil.html

Amazon rainforest will be halved within 20 years.
  http://news.mongabay.com/2008/0227-nepstad_amazon.html

Amazon rainforest the size of Western Europe
  http://rainforests.mongabay.com/amazon/external_nov04.html

Brazil is world's second largest producer of soybeans.
  http://www.ers.usda.gov/Briefing/SoyBeansOilCrops/2007baseline.htm

Area harvested for soybeans in Brazil (as of 2005) equal to 53.1 million acres (83,000 square miles)
  http://www.agriculture.com/ag/story.jhtml?storyid=/templatedata/ag/story/
    data/1133211356830.xml&catref=ag1001

Size of the state of Kansas is 81,814 square miles
  http://quickfacts.census.gov/qfd/states/20000.html

Soy oil most common edible oil
  http://www.bothends.org/strategic/res_palmoil_wwf3.pdf

Soybeans and McDonalds
  http://www.nature.org/magazine/autumn2007/features/art21918.html

Nearly 80 percent of global soybean harvests are milled into animal feed.
  http://www.nature.org/magazine/autumn2007/features/art21918.html

Percentage of biodiesel in US and Brazil derived from soy
  http://www.worldwatch.org/node/5442

"Hamburger Connection Fuels Amazon Destruction," CIFOR
  http://www.cifor.cgiar.org/publications/pdf_files/media/Amazon.pdf

Kaimowitz: "Cattle ranchers are making mincemeat out of Brazil's Amazon rainforests"
  http://www.cifor.cgiar.org/PressRoom/MediaRelease/2004/2004
    _04_02.htm

Brazil world's largest meat exporter with 25 percent of world market share
  http://www.fao.org/docrep/010/ah864e/ah864e09.htm

Pounds of beef per acre of land
  http://www.beeffrompasturetoplate.org/mythmeatproductioniswasteful.
    aspx#Just%20165%20pound

Amazon tree density up to 300 trees per acre
  http://earth.leeds.ac.uk/ebi/publications/Steege_2003.pdf
  http://www.fs.fed.us/r5/sierra/projects/environassess/kingsriver/2006-Final/
    apna-historical.pdf

Forest products rank third among Brazil's agricultural exports.
http://www.fas.usda.gov/gainfiles/200612/146269769.pdf

Percent of tropical wood from predatory sources; logged illegally; certified logging
http://www.fas.usda.gov/gainfiles/200612/146269769.pdf

US third largest consumer of Brazilian topical lumber and top consumer of Brazilian softwood
http://www.fas.usda.gov/gainfiles/200612/146269769.pdf

Uses of eucalyptus and pine imported to US
http://www.worldforestinvestment.com/southern.pdf

Brazil imports rubber from Malaysia.
http://www.statistics.gov.my/english/frameset_rubber.php?file=rubberJan08

Story of Henry Alexander Wickham
http://www.irrdb.com/IRRDB/NaturalRubber/History/HenryWickham.htm

Southeast Asia produces more than 70 percent of global rubber supply.
http://www.deltaplanttechnologies.com/pr_100206.asp

AP quote regarding Brazil's environmental laws
http://climate.weather.com/articles/amazon122407.html

Land titles in Brazil
http://www.marcusmarsiglia.com/brazil-real-estate.html

Petrobas is Brazil's largest energy company and the eighth biggest oil company in the world
http://www2.petrobras.com.br/ri/ing/ConhecaPetrobras/ConhecaPetrobras.asp

The population of Manaus grew by more that 65 percent between 1993 and 2003 to its current population of over 1.5 million.
http://new.unep.org/Documents.Multilingual/Default.asp?DocumentID=487&ArticleID=5350&l=en

Land use in the US
http://www.ers.usda.gov/publications/EIB14/eib14a.pdf

Brazil farms on less than half the land cropped by the US, yet has 50 percent more potential cropland.
http://www.agri-industries.com/articles/BrazilCrop.pdf

Total GDP of Brazil vs. US
https://www.cia.gov/library/publications/the-world-factbook/fields/2195.html

London Telegraph quote on REDD
http://www.telegraph.co.uk/earth/main.jhtml?xml=/earth/2007/12/14/eacanopy114.xml

Pilot program to pay forest-dwelling families for not logging or burning forest
  *http://www.telegraph.co.uk/earth/main.jhtml?xml=/earth/2007/12/14/
  eabali214.xml*

Online carbon calculators
  http://www.carbonfund.org/site/pages/carbon_calculators/
  http://www.epa.gov/climatechange/emissions/ind_calculator.html
  http://www.safeclimate.net/calculator/

Time to grow trees for pulpwood
  http://www.fao.org/docrep/f7795e/f7795e01.htm

Per capita consumption of paper and paperboard (2005 figure) is 655 pounds per year (1.8 pounds per day). At 100 sheets per pound, this is 1260 sheets per week and 65,500 sheets per year.
  http://earthtrends.wri.org/
  http://eetd.lbl.gov/paper/ideas/html/copyfactsA.htm

One tree makes 8,333 sheets of copy paper
  http://www.conservatree.com/learn/EnviroIssues/TreeStats.shtml

Six hundred billion dollars of natural resources
  http://www.american.edu/TED/brazil.htm

CHAPTER 7: THE REALITY OF OUR ACTIONS

Verrazano-Narrows is the longest suspension bridge in the US
  http://www.nycroads.com/crossings/verrazano-narrows/

Fresh Kills landfill could be considered the largest man-made structure on earth, visible from space with the naked eye.
  *http://www.waste-management-world.com/display_article/314941/123/
  CRTIS/none/none/1/Fresh-Kills/*

Fresh Kills size statistics
  http://www.nyc.gov/html/dcp/html/fkl/fkl6.shtml

Highest point on the eastern seaboard
  http://www.mcdonough.com/writings/new_geography.htm

World Trade Center wreckage sent to Fresh Kills
  http://www.nysm.nysed.gov/exhibits/longterm/documents/recovery.pdf

The basics of landfills
  http://www.epa.gov/epaoswer/non-hw/muncpl/landfill/sw_landfill.htm

Sources and types of trash
  http://www.epa.gov/epaoswer/non-hw/muncpl/pubs/msw06.pdf

Paper recycling rate figure calculated by multiplying total amount of paper (minus paperboard) recycled (roughly 20 million tons) by number of reams per ton (estimated at 5 pounds per ream, or 400 reams per ton), which equals the equivalent of 7.9 billion reams recycled per year. With about 150 million females in the US, this is 52.5 reams recycled per female per year, or about one per week.

http://www.epa.gov/epaoswer/non-hw/muncpl/pubs/06data.pdf
www.census.gov/popest/national/asrh/NC-EST2005/NC-EST2005-01.xls

Plastic-recycling-rate figure calculated by multiplying total amount of plastic containers recycled (roughly 3.02 billion pounds) by the number of two-liter bottles per pound (estimated at 1/8 pound per bottle, or 8 bottles per pound), which equals the equivalent of 24.2 billion two-liter bottles recycled per year. With about 72 million Americans living today who were born between 1946 and 1964 (Baby Boomer generation), this is 336 two-liters per Baby Boomer—or nearly one per day.

http://www.epa.gov/epaoswer/non-hw/muncpl/pubs/06data.pdf
http://www.epa.gov/epaoswer/non-hw/recycle/recmeas/docs/guide_b.pdf
www.census.gov/popest/national/asrh/NC-EST2005/NC-EST2005-01.xls

Glass-recycling rate figure calculated by multiplying total amount of glass containers recycled (roughly 5.76 billion pounds) by the number of glass bottles per pound (about 2 bottles per pound), which equals the equivalent of 11.5 billion glass bottles recycled per year. With about 60.7 million Americans under the age of 15, this is 190 glass bottles per kid—or about one for each day of the school year.

http://www.epa.gov/epaoswer/non-hw/muncpl/pubs/06data.pdf
http://www.epa.gov/epaoswer/non-hw/recycle/recmeas/docs/guide_b.pdf
www.census.gov/popest/national/asrh/NC-EST2005/NC-EST2005-01.xls

Per-capita waste generation from 1960 to 2006
http://www.epa.gov/epaoswer/non-hw/muncpl/pubs/msw06.pdf

Population estimates for 1960 and 2008
http://www2.census.gov/prod2/decennial/documents/09768103v1p1_TOC.pdf
http://www.census.gov/

Recycling rates from 1960 to 2006
http://www.epa.gov/epaoswer/non-hw/muncpl/pubs/msw06.pdf

Number of landfills in operation in 1988 vs. 2006
http://www.epa.gov/msw/facts.htm

Landfill size increase in last decade from 1 million tons to 25 million tons.
http://www.cabq.gov/solidwaste/overview/cerro

Americans generate 251 million tons of waste per year and recycle 82 million tons.
http://www.epa.gov/epaoswer/non-hw/muncpl/pubs/msw06.pdf

EPA campaign to achieve 25 percent recycling and source reduction rates
http://www.epa.gov/epaoswer/general/k02027.pdf

Eighty-one million plastic water bottles tossed per day. Figure calculated by taking annual number of water bottle units purchased (29.8 billion as of 2005) and dividing by 365 days per year.
http://www.container-recycling.org/assets/pdfs/reports/2007-waterwater.pdf

More than a billion pounds of plastic saved from landfills each year. Figure calculated by assuming 50 percent fewer of the 29.8 billion water bottles sold would be sent to landfills each year if every water bottle were used twice before being discarded. Given 12 bottles per pound and an estimated landfill/incineration/littering rate for water bottles of 85.5 percent, roughly 1.03 billion pounds of plastic could be saved from landfills if each water bottle were refilled before being trashed or recycled.
http://container-recycling.org/assets/pdfs/reports/2007-waterwater.pdf

One hundred billion plastic grocery bags per year
http://www.worldwatch.org/node/1499

Metals recycling eliminates greenhouse gas emissions.
http://www.epa.gov/epaoswer/non-hw/muncpl/pubs/msw06.pdf

EPA quote about benefits of recycling
http://www.epa.gov/epaoswer/non-hw/muncpl/pubs/msw06.pdf

If there were no recycling, an estimated 2.3 billion tons of materials—computed from data for every five years since 1960—would have to be dealt with.
http://www.epa.gov/epaoswer/non-hw/muncpl/pubs/msw06.pdf

Seventy-five percent of the land space we currently use for trash if we only recycled more. In the US, that land space is the size of Pennsylvania
Elizabeth Rogers, Thomas M. Kostigen. 2007. *The Green Book*. New York: Three Rivers Press.

Two billion three hundred million tons of trash compressed at a density of 250 pounds per cubic yard would cover 12.4 billion square yards (4,000 square miles) to a height of almost 4.3 feet. LA County covers an area of 4,060 square miles.
http://quickfacts.census.gov/qfd/states/06/06037.html

If we didn't recycle the 20 million tons of paper each year, we'd need an additional 340 million trees (17 trees per ton of paper), 40 million bar-

rels of oil (2 barrels per ton), and 82 billion kilowatt-hours of electricity (4,100 kWh per ton).
http://www.epa.gov/rcc/onthego/benefits/index.htm#paper

If we didn't recycle the 690,000 tons of aluminum, we'd need an additional 35 billion kilowatt-hours of electricity. An average coal-fired power plant produces 3.5 billion kWh per year. Four tons of bauxite ore are needed per ton of aluminum.
http://container-recycling.org/alum_rates.htm
http://www.ucsusa.org/clean_energy/fossil_fuels/offmen-how-coal-works.html
http://www.epa.gov/sectors/pdf/energy/ch3-1.pdf

Four mountains made from 150 million tons of solid waste
http://www.nycgovparks.org/sub_your_park/fresh_kills_park/html/history.html

Incineration/combustion rate in the US
http://www.epa.gov/epaoswer/non-hw/muncpl/pubs/msw06.pdf

Khian Sea barge
http://highered.mcgraw-hill.com/sites/0072919833/student_view0/
chapter13/additional_case_studies.html
http://query.nytimes.com/gst/fullpage.html?res=9F0CE3DD1330F934
A35753C1A965958260

Forty billion dollar waste market
http://wasteage.com/research/

Two hundred fifty billion dollar recycling market
http://www.wbcsd.org/plugins/DocSearch/details.asp?MenuId=Nzk&Click
Menu=&doOpen=1&type=DocDet&ObjectId=MjgwMzg

Waste into space, oceans, snow
http://space.newscientist.com/article/mg18625012.700-dump-it-in-the-
mantle.html

History of Fresh Kills
http://www.nycgovparks.org/sub_your_park/fresh_kills_park/html/history.html

Average daily tonnage received by California landfills
www.ciwmb.ca.gov/agendas/mtgdocs/2002/02/00007306.doc

Anaerobic decomposition of landfill waste tied to greenhouse gases
http://www.ecocycle.org/TimesSpring2002/NewEvidence.cfm

Time it takes for garbage to decompose
http://www.des.state.nh.us/Coastal/Trash/documents/marine_debris.pdf

Landfills are the largest emitters of methane in the country.
http://uspowerpartners.org/Topics/SECTION6Topic-LandfillMethane.htm

Recycled content vs. postconsumer recycled
http://earth911.org/recycling/identifying-recycled-content-products/

Fresh Kills is responsible for 5.7 percent of US methane emissions and 1.8 percent of global emissions, according to Nick Dmytryszyn, the environmental engineer for the borough.
http://query.nytimes.com/gst/fullpage.html?res=9C00E2DA1F3EF934 A15757C0A961958260

Air pollution from Fresh Kills
http://www.atsdr.cdc.gov/hac/pha/freshkills/fkl_p2.html

Incinerator health hazards
http://www.mindfully.org/Health/Incinerator-Childhood-Cancers2000.htm

Battery recycling
http://www.batteryrecycling.com/experts.html

Plastics recycling codes
http://www.valcorerecycling.org/affair/archives/2002-08-04.htm

Average landfill disposal fee for one ton of trash
http://wasteage.com/mag/waste_bargain_basement_tipping/

Only 7 percent of plastics are recycled.
http://www.epa.gov/epaoswer/non-hw/muncpl/pubs/06data.pdf

Waste hauling and transfer stations
http://wasteage.com/mag/waste_transfer_station_transition/

Dumps vs. landfills
http://people.howstuffworks.com/landfill3.htm

Landfill maintenance and leachate collection
http://people.howstuffworks.com/landfill6.htm

Methane capture from landfills
http://www.p2pays.org/ref/11/10985.pdf

1995 Fresh Kills health hazards study
http://www.atsdr.cdc.gov/hac/pha/freshkills/fkl_p2.html

Recycling statistics
http://www.nrc-recycle.org/theconversionator/shell.html

Landfill health effects
http://www.gfredlee.com/cal_risk.htm

Fresh Kills Park Project
http://www.nycgovparks.org/sub_your_park/fresh_kills_park/html/fresh_ kills_park.html

CHAPTER 8: WHERE THE CURRENTS TAKE OUR TRASH

Annual production of plastic
http://www.algalita.org/pdf/AMRFWhitePaper.pdf

Proportion of ocean debris dumped from sea vessels vs. from shore
http://www.publicaffairs.noaa.gov/oceanreport/marinedebris.html

Oceans occupy 70 percent of the earth's surface and are home to more than 90 percent of the organisms on the planet.
http://marinebio.org/MarineBio/

Worldwide dependence on fisheries
http://www.idrc.ca/en/ev-28113-201-1-DO_TOPIC.html

Quantity of marine debris per square mile of the Eastern Garbage Patch
http://www.algalita.org/pdf/Action-sheet.pdf

Rubbish superhighway to garbage patch in Asia
http://www.abc.net.au/news/stories/2007/11/01/2079239.htm?section=world

One hundred twenty-nine million plastic beverage bottles discarded everyday
http://container-recycling.org/images/plastic/graphs/PETwastesale-unit-96-06.gif

Concentration of plastic vs. zooplankton
http://www.alguita.com/pdf/Density-of-Particles.pdf

Blue whale diet
http://animals.nationalgeographic.com/animals/mammals/blue-whale.html

"Marine debris collects within the North Pacific Subtropical Convergence Zone." *Marine Pollution Bulletin,* Volume 54, Issue 8, August 2007, pp. 1207–1211
www.elsevier.com/locate/marpolbul (subscription required)

Health threats from ingesting certain chemicals found in plastics
http://www.atsdr.cdc.gov/tfacts68.pdf
http://www.jsonline.com/story/index.aspx?id=710303

Forty-six thousand pieces of plastic litter per square mile of ocean
http://www.unep.org/Documents.Multilingual/Default.asp?DocumentID=480&ArticleID=5300&l=en

Seventy percent of marine litter sinks
http://marine-litter.gpa.unep.org/facts/facts.htm

Walt Whitman, *Leaves of Grass* (1860)
http://www.whitmanarchive.org/published/LG/1860/poems/27

Phytoplankton responsible for half of the world's photosynthesis
http://oceancolor.gsfc.nasa.gov/SeaWiFS/TEACHERS/BIOLOGY/

Ocean carbon cycle
http://science.hq.nasa.gov/oceans/system/carbon.html

One hundred-year time horizon
 "Should oceanographers pump iron?" *Science,* Vol. 318, pp. 1368–70,
 November 2007
 http://www.sciencemag.org/cgi/content/summary/318/5855/1368
 (subscription required).

Pieces of plastic covering the seafloor
 *http://oceans.greenpeace.org/raw/content/en/documents-reports/plastic_
 ocean_report.pdf*

Northwestern Hawaiian Islands National Monument
 http://www.whitehouse.gov/news/releases/2006/06/20060615-6.html

Marine animal (albatross, whale, dolphin, sea turtle) deaths from bycatch
 http://www.pifsc.noaa.gov/library/bycatch.php

Scientists have identified 267 types of marine species killed by ocean debris.
 http://www.healthebay.org/currentissues/ppi/marinedebris.asp

Plastic debris found in guts of Antarctica's snow petrel chicks
 *http://oceans.greenpeace.org/raw/content/en/documents-reports/plastic_
 ocean_report.pdf*

Mississippi River drains 41 percent of the United States
 http://water.usgs.gov/nasqan/progdocs/factsheets/missfact/missfs.html

Eight million pieces of marine litter enter oceans every day.
 http://www.unep.org/regionalseas/marinelitter/about/distribution/default.asp

International Coastal Cleanup Day statistics
 *http://www.oceanconservancy.org/site/DocServer/2007ICCFactSheet.
 pdf?docID=2842*

Ocean currents
 http://earth.usc.edu/~stott/Catalina/Oceans.html

Deep water circulation
 http://earth.usc.edu/~stott/Catalina/Deepwater.html

Global conveyor belt
 http://www.columbia.edu/cu/record/23/11/13.html

Upwelling
 *http://www.nwfsc.noaa.gov/research/divisions/fed/oeip/db-coastal-
 upwelling-index.cfm*

The figure of 1,000 pieces of plastic per person in the world was calculated
 by multiplying 46,000 pieces of marine litter by 139 million square miles
 of ocean, and dividing by 6.6 billion people.
 http://www.eoearth.org/article/Ocean

Pacific Ocean facts
 http://www.eoearth.org/article/Ocean

The figure of almost 3 trillion pieces of plastic floating in the Pacific Ocean alone was calculated by multiplying 46,000 pieces of marine litter by 60 million square miles of Pacific Ocean.

The Great Los Angeles River Cleanup
  http://www.folar.org/programs.html

The link between grease and sewage spills
  http://www.stockteam.com/hbpress7.html
  http://www.sierravistawater.com/docs/fog.pdf

Algalita blog
  http://orvalguita.blogspot.com/

Mercury in seafood
  http://www.epa.gov/waterscience/fish/advice/index.html

CHAPTER 9: THE GREATEST PROBLEM NO ONE HAS HEARD ABOUT

Survival without water
  http://www.med-library.net/content/view/324/41/

Total global water withdrawals in 2025 compared to current rate
  http://www.ifpri.org/pubs/fpr/fprwater2025.pdf

Individual indoor water use in the US
  http://www.aquacraft.com/Publications/resident.htm

Cost to upgrade water infrastructure in US and worldwide
  *http://www.strategy-business.com/press/article/07104?gko=a8c38-1876
  -23502998*

Clean Water Act
  http://www.epa.gov/region5/water/pdf/ecwa.pdf
  Quote from PhD water experts Bill Jury and Henry Vaux Jr. from "The Emerging Global Water Crisis: Managing Scarcity and Conflict Between Water Users" *Advances in Agronomy*, vol. 95, Oct. 2007.

Number of people worldwide lacking access to safe drinking water
  http://www.worldwatercouncil.org/index.php?id=25

Estimate of the number of people who die from a cause related to lack of clean water
  *http://www.pacinst.org/reports/water_related_deaths/water_related_
  deaths_report.pdf*

Water use and relative water prices around the world
  http://mediaglobal.org/index.php/fast-facts/

Climate change and global water scarcity
  http://www.worldwatch.org/node/1779

Climate change and severe drought occurrence
http://earthtrends.wri.org/updates/node/264

Himalayan glacier melt
Maude Barlow. 2007. *Blue Covenant.* New York: The New Press.

China population and water supply
http://www.worldwatch.org/node/3893

Nine countries possess 60 percent of the world's water supply.
http://www.financialsense.com/editorials/dickerson/2007/0308.pdf

Countries of the world
http://www.state.gov/s/inr/rls/4250.htm

Water used by agriculture worldwide and in developing countries
http://earthtrends.wri.org/updates/node/264

Water requirement to satisfy daily food intake
http://www.worldwatercouncil.org/index.php?id=866

Water supply 17 percent short of amount needed to feed 2020 global population, according to World Water Council
http://news.bbc.co.uk/1/hi/sci/tech/755497.stm

World's top three grain producers
http://www.earth-policy.org/Indicators/Grain/2006.htm

Virtual water
http://www.worldwatercouncil.org/fileadmin/wwc/Programs/Virtual_Water/virtual_water_final_synthesis.pdf

Water requirements for rice (370 gal.) and beef (3,434 gal.)
http://www.worldwatercouncil.org/index.php?id=25

World Water Council quote
http://www.worldwatercouncil.org/index.php?id=866

The Great Lakes
http://www.epa.gov/grtlakes/

Plan to ship Lake Superior water in tankers to Asia
http://www.vjel.org/journal/VJEL10044.html

Harper's magazine (September 2007) article
http://www.harpers.org/archive/2007/09/0081685 (subscription required)

At least 36 states will face water shortages within five years.
http://www.epa.gov/WaterSense/water/why.htm

Christian Science Monitor article and Bill Richardson Las Vegas statements about water policy
http://www.csmonitor.com/2008/0107/p03s03-uspo.html

Georgia water crisis
  http://abcnews.go.com/GMA/story?id=3730145

Water rights in the US
  http://www.waterencyclopedia.com/La-Mi/Law-Water.html

More than 30 million people rely on the Great Lakes for drinking water.
  http://www.epa.gov/grtlakes

Great Lakes water levels reached record low in 2007.
  *http://www.usatoday.com/news/nation/environment/2007-06-13-lake-
  superior_N.htm*
  *http://www.lre.usace.army.mil/_kd/Items/actions.
  cfm?action=Show&item_id=3886&destination=ShowItem*

Rising water temperatures of the Lakes since 1980
  *http://www.mlive.com/news/bctimes/index.ssf?/base/
  news-11/1199981750220130.xml&coll=4*

Dasani source
  http://www.dasani.com/flash.htm

One quarter of bottled water is tap water.
  http://www.nrdc.org/water/drinking/qbw.asp

Tap water percent of total fresh water use
  http://ga.water.usgs.gov/edu/wups.html

Water allocation in the US
  http://esa21.kennesaw.edu/activities/water-use/water-use-overview-epa.pdf

The water cycle
  http://earthobservatory.nasa.gov/Library/Water/water_2.html

The science of thermal expansion
  *http://www.watts.com/pro/divisions/watersafety_flowcontrol/learnabout/
  learnabout_thermexpansion.asp#generalinfo*

Sea-level rise due to thermal expansion and melting glaciers and ice sheets
  *http://earthobservatory.nasa.gov/Newsroom/NasaNews/2006
  /2006061422488.html*

Desalination
  http://ga.water.usgs.gov/edu/drinkseawater.html

Toilet-to-tap in California
  "O.C. sewage will soon be drinking water." *LA Times,* Jan 3, 2008

A household of four could save 14,000 gallons of water each year with low-
  flush toilets, 1,700 gallons with low-flow faucet aerators, and 20,000
  gallons per year or more with low-flow showerheads.
  http://www.h2ouse.org/tour/bath.cfm

Turning the water off while brushing your teeth saves an estimated 2.6 gallons per brushing (assuming 1.3 gallons per minute and two minutes of brushing). Assuming two brushings per day, this is 1,423 gallons per year.
http://www.h2ouse.org/tour/bath.cfm

Toilet leak dye test
http://www.h2ouse.org/tour/bath.cfm

Dishwashers use an average of 9 gallons per load while hand washing uses an average of 20 gallons per load.
http://www.energystar.gov/ia/partners/manuf_res/downloads/
2007Dishwasher_prg.pdf

Sixty-gallon bathtubs vs. 10-minute showers (at 2.5 gallons per minute)
http://www.fypower.org/res/tools/energy_tips_results.
html?tips=water-heating

Plant water-conserving plants
http://www.h2ouse.org/gardensoft/index.aspx

Eliminate one flush per day.
http://www.h2ouse.org/tour/bath.cfm

Commercial car washes vs. hand washing
http://planetgreen.discovery.com/transport-tech/vroom-vroom/car-wash.php

An average of 400 gallons of water is saved each time one-quarter pound of beef is replaced with one-quarter pound of soy.
http://www.waterfootprint.org/Reports/Report12.pdf

Percentage of indoor water use flushed down the toilet
http://www.h2ouse.org/tour/bath.cfm

Chapter 10: Where to Spread Your Wings

Population of US residents living in Pacific Time Zone calculated from summing 2006 population estimates of California, Nevada (minus populations of Jackpot and West Wendover, which are on Mountain Standard Time), Oregon (minus four-fifths of the population of Malheur County, which is on Mountain Standard Time), Washington, and the ten counties that make up the Idaho panhandle.
http://quickfacts.census.gov/qfd/index.html
http://www.access.gpo.gov/nara/cfr/waisidx_03/49cfr71_03.html

US population estimate as of March 2008 is more than 303 million.
http://www.census.gov/main/www/popclock.html

US spends $1 million per minute on energy.
www.eia.doe.gov/kids/energyfacts/saving/efficiency/savingenergy.html

According to some estimates, daylight saving time may trim the nation's energy bill by one percent per day.
http://nationalatlas.gov/articles/boundaries/a_savingtime.html

Total energy expenditures (residential, commercial, industrial, transportation) in the US in 2005 exceeded $1.04 trillion.
http://www.eia.doe.gov/oiaf/archive/aeo07/pdf/appendixes.pdf

The ideas, history, and arguments behind daylight saving time
http://webexhibits.org/daylightsaving/c.html

Daylight saving time controversy in recent research from the *Wall Street Journal*
http://online.wsj.com/public/article/SB120406767043794825.html

Allocation of energy consumed by US office buildings
http://www.eere.energy.gov/buildings/info/office/

Average square footage of new home in US
http://www.nahb.org/news_details.aspx?newsID=4143

Average home energy usage and energy expenditures
http://www.eia.doe.gov/emeu/recs/recs2001/ce_pdf/enduse/ce1-61u_hhmem_useind2001.pdf

Average home water use (household of four)
http://www.aquacraft.com/Publications/resident.htm

Average home waste generation (household of four)
http://www.epa.gov/msw/facts.htm

Average water use for households in the Metropolitan Water District of Southern California
http://www.mwdh2o.com/mwdh2o/pages/yourwater/story/story01.html
For water information around California visit *http://www.water-ed.org/watersources/default.asp*

Metropolitan Water District explanation about process of water treatment
http://www.mwdh2o.com/mwdh2o/pages/yourwater/story/story01.html

The fate of household wastewater in Los Angeles
http://www.lasewers.org/treatment_plants/hyperion/index.htm

The Gap organic cotton tee
http://gapinc.com/public/Media/Press_Releases/med_pr_OrganicT032107.shtml

Levi's organic cotton jeans
http://www.levistrauss.com/News/PressReleaseDetail.aspx?pid=784

Worn Again recycled shoes
www.wornagain.co.uk

Country of origin for textile and clothing imports to US
www.ers.usda.gov/Data/FiberTextileTrade/chartstables/Figure4.xls

Recent growth in clothing imports from Vietnam
http://www.ers.usda.gov/data/fibertextiletrade/

Honduran socks on National Public Radio
http://www.npr.org/templates/story/story.php?storyId=16661333
http://www.npr.org/templates/story/story.php?storyId=16673310

Percentage of apparel purchased in the US made domestically, apparel expenditures, and garments purchased annually.
http://www.apparelandfootwear.org/UserFiles/File/Statistics/trends2005.pdf

Plastics consumption and waste
http://www.epa.gov/epaoswer/non-hw/muncpl/pubs/msw06.pdf

Ten cents of every dollar spent pays for packaging.
http://www.wastediversion.org/residentialrecycling.htm

Manufacturing energy consumption
http://www.eia.doe.gov/emeu/aer/consump.html

Industrial water use
http://pubs.usgs.gov/circ/2004/circ1268/htdocs/text-total.html

Industrial waste
http://www.epa.gov/industrialwaste/

One trillion plastic bags used per year
http://www.reusablebags.com/facts.php

Packaging comprises most waste at landfills.
http://www.epa.gov/epaoswer/non-hw/muncpl/pubs/msw06.pdf

Miles food travels from farm to table
http://www.worldwatch.org/node/1749

Most cardboard fiber recycled in the US is reprocessed in China and sold back to the US according to Patrick J. Moore, chairman and CEO of the Smurfit-Stone Container Corporation.
http://www.nytimes.com/2007/12/15/business/15interview.html

Recycling rate in Santa Monica
http://www.smdp.com/article/articles/2512/1/The-rigors-of-recycling/Page1.html/print/2512

National recycling rate
http://www.nrc-recycle.org/gaorecyclingreport.aspx

Southern CA energy sources

http://www.indiacore.com/bulletin/papers-di1/bharat-bhargava-brief-overview-of-southern-california-edison-power-system-in-united-states-of-america.pdf

Phantom power

http://www.des.state.nh.us/gw/gw0906.htm

CFLs

http://www.fypower.org/res/tools/products_results.html?id=100195

Index

acid rain: environmental impact of, 12–13; UNESCO-designated World Heritage Sites damaged by, 14

Agarwal, Ravi, 28

air pollution: AirNow.gov developed to educate about, 107; "Coriolis effect" of, 80; efforts to reduce, 69–70, 71; estimated annual deaths from, 106; global measurements of, 107; health and economic consequences of, 71, 72; impact on children by, 72; from landfill decomposition, 131–32; Linfen (China), 61–62, 63–65; US consumption contributing to global, 64–67; Western Wall (Jerusalem) damage due to, 11. *See also* pollution

air quality: best and worst areas of US, 107; calculations for air pollutants by EPA, 107

AirNow.gov, 107

Alaska Climate Impact Assessment Commission, 87

albedo effect, 96–97

Algalita Marine Research Foundation, 147–49, 152–53, 196

Alguita blog, 162–63, 203

Alguita (ship), 145, 147, 162, 196

Allan Company recycling center (Santa Monica), 133

Amazon Conservation Team, 115

Amazon rainforest: animal species of the, 104–5, 116–17; corporate displacement of indigenous people of, 112–13; deforestation of the, 103–4, 107–12; as lungs of the Earth, 103–4; multiple natural resources and global benefits of, 116–18; possible solutions to deforestation of, 120–21; soybean production resulting in destruction of, 108–9. *See also* Brazil

animal species: Amazon rainforest, 104–5, 116–17; of Borneo (Southeast Asia), 43–44, 50; environmental costs of banishing, 58–59; global identification of, 58; ocean pollution impact on, 155

Arava Institute (Israel), 15–16

Arctic haze, 97

Arctic pollution: dirty snow consequence of, 92–93; sources of and materials included as part of, 93–96

Army Corps of Engineers, 82, 91

Attenborough, David, 119

Avila, Don, 135, 136

BAN (Basel Action Network), 28–29, 30

Bangladesh: cyclone destruction in, 52–53; mangrove forest of Sundarbans National Park in, 53

bauxite mining, 129–30

beef exports (Brazil), 109–10

Beijing Olympics (2008), 66

Borneo (Southeast Asia): bleak future of orangutans of, 50; environmental costs of deforestation of, 58–59; exploring the jungle of, 51–52; geography and plant and animal species of, 43–44; logging camps of, 41–42, 45, 55; palm oil plantation deforestation of, 43–45, 46–50, 55–57; WWF-Malaysia approach to sustainability in, 45–47, 48–49, 56. *See also* Malaysia

Braga, Eduardo, 119

Brazil: corporate displacement of indigenous people of, 112–13; declining rubber market of, 110–11; demand for beef exports from, 109–10; growth of Manaus area in, 113–14; land and agricultural distribution in, 116; logging industry in, 110–11; natural resources and "here and now" mentality in, 114–15; soybean industry of, 108–9; strict environmental laws of, 111–12. *See also* Amazon rainforest

British Environmental Protection Agency, 27

Brunei (Malaysia), 46

Bush, George W., 154–55, 198

Cahill, Cathy, 93, 94

Canadian Clean Power Coalition, 75–76

Canadian Climate Center, 23

carbon credits, 119

carbon dioxide emissions: annually produced by every person in the US, 94; burning coal producing, 76; deforestation accounting for, 46; greenhouse gases made up of, 11; Kyoto Protocol (1997) on, 69–70; ocean pollution producing, 153–54; simple actions by individuals to cut, 100–101; US consumption contributing to, 64–67

carbon monoxide, 107

carpooling, 100

Catherine of Braganza, 26

Center for Biology of Natural Systems (Queens College), 95

Chan, Jackie, 63

Charles II (England), 26

children health issues, 72

China: air pollution problem in, 61–80; Guangdong province polluter in, 66; insufficient water supply of, 172; Lake Tai contamination in, 68; lead-tainted toys from, 67; UNESCO-designated World Heritage Sites damaged, 14. *See also* Linfen (China)

The China Daily, 74

Christian Science Monitor, 176

CIFOR (Center for International Forestry Research) [Indonesia], 109

Clean Air Act, 107

clean coal power, 77–78
clean MRFs (material recovery facilities), 135–36
Clean Water Act, 168
climate change: albedo effect contributing to, 96–97; economic damage resulting from, 71; faced by Mumbai (India), 23–26; impact on water supply by, 171–72; intangible damages due to, 14–15; Ken Stenek's teaching curriculum on, 98–100; as leading cause of destruction, 13–14; NOAA reporting on winter of 2007 and, 92; Shishmaref Village (Alaska) experiencing impact of, 81–101; widespread flooding due to, 14. *See also* global warming
Climate Prediction Center, 17
climate versus weather, 18
clothing imports, 197–98
coal power: as Arctic pollution source, 94–95; Canadian Clean Power Coalition on process of, 75–76; clean coal, 77–78; Linfen production/air pollution and, 61–62, 63–65, 72–74, 79–80; pollution resulting from, 12, 73, 75, 76–77; US use of, 77
coastal regions: Bangladesh mangrove forest as natural protection of, 53; climate change and vulnerability of, 23–24; melting glaciers and ice caps affecting, 97–98; Mumbai example of problems associated with, 22–25; United States population along, 22, 97–98

coca leaves, 117
"Commando" weather report (Weather Channel), 156
Computer Village (Nigeria), 28
Conservation Action Trust (India), 53
conservation measures: to cut carbon emissions, 100–101; environmentally-friendly clothing as, 196–99; to extend water supply, 186–88, 194–96; importance of taking individual, 207–9; to reduce ocean pollution, 161–62; regarding food packaging and products, 199–200; to save on electricity/ energy, 100, 193–94, 204–6. *See also* environmental activism
"Coriolis effect," 80

Dayan, Uri, 13
daylight saving time, 191–93
deforestation: of the Amazon rainforest, 103–4, 107–12; Borneo's palm oil production and, 43–45, 46–50, 55–57; environmental costs of, 58–59; global carbon dioxide emission produced by, 46; of mangrove forest (Bangladesh), 53; paper recycling decreasing, 129; possible solutions to Amazon rainforest, 120–21; witnessing process of, 54–55. *See also* logging industry
Deonar landfill (Mumbai, India), 31–32
desalination, 184
Dharavi Redevelopment Project, 39

Dharavi slum (Mumbai, India), 33–40

Dickerson, John, 170, 172, 176, 186

dioxin emissions: description and rising levels of, 95; health problems caused by, 95–96

dirty MRFs (material recovery facilities), 135–36

dirty snow, 92–93

Dome of the Rock (Jerusalem), 6

drinking water supply: declining access to safe, 169–74, 175–76; individual actions to extend, 186–88, 194–96; Lake Tai contamination (China) of, 68; measures to expand, 184–86; Mumbai (India) lack of, 22–23. *See also* freshwater supply

Duluth (Minnesota), 179–81, 188–89

dumps, 137

Earth: Amazon rainforest as lungs of the, 103–4; connections to everything in the, 18–19; importance of caring for the, 207–9; water cycle of, 182–84

Eastern Garbage Patch (Pacific Ocean): Algalita Marine Research Foundation study of, 147–49, 152–53, 196; description of, 143, 146–47; sources of and environmental impact of, 144–49, 153–60; taking a tour of, 149–52. *See also* Western Garbage Patch (Pacific Ocean)

Ecce Homo convent of the Sisters of Sion (Jerusalem), 6

"An Economical Project" (Franklin), 192

Edison, Thomas, 76

El Niño, 17

electricity: Edison's coal-fired production of, 76; garbage-burning to produce, 78; taking measures to save on, 100, 193–94, 204–6

Ellesmere Island (Canada), 92

energy consumption: of average US home, 193; consumer contributions to improving, 78–79; daylight saving time to conserve, 191–93; economic cost of US, 191; efforts to conserve home, 100, 193–94, 204–6

energy infrastructure: coal power component of, 72–73, 77–78; efforts to find other energy sources for, 108; need to rebuild our, 78

environmental activism: importance of individual awareness and, 207–9; by individuals to extend water supply, 186–88; palm oil production condemned by, 44–45; three Rs (reduce, reuse, and recycle) of, 139. *See also* conservation measures

Environmental Protection Agency (EPA): national air quality standards established by, 107; national recycling campaign of, 126, 127, 128; on reducing smog levels, 66; Sarichef Island report made to, 91

environmentally-friendly clothing, 196–99

Eriksen, Marcus, 149, 151, 155–56
ethanol, 108, 113
e-waste: BAN (Basel Action Network) study of, 28–29; Computer Village (Nigeria) landfill site of, 28; global generation of, 27–28; Indian problem with, 27–30; movement to ban exports of, 30; Silicon Valley Toxics Coalition report on, 29; toxic dust produced by, 37; Toxics Link tracking, 27; US production of, 33. *See also* hazardous waste; landfills

FDA (Food and Drug Administration), 164
flooding: environmental damage due to, 14; melting glaciers and ice caps resulting in, 97–98; Mumbai (India), 22, 23
food packaging, 199–201
Forest Stewardship Council (FSC), 121
fossil fuels, 11
Franklin, Benjamin, 192
Fresh Kills landfill (New York): attempts to transform legacy of, 140–41; description of, 124–25, 136–39, 141; massive size of, 123–24, 130, 131, 132; methane gas emission of, 132, 138; opening, use of, and lawsuits over, 130–31
Fresh Kills Park Project, 140–41
freshwater supply: desalination of, 184; recycling wastewater to expand, 185–86. *See also* drinking water supply

Gandhi, Mahatma, 26
Gateway of India (Taj Mahal Hotel), 26
Global Canopy Programme, 118
global warming: accelerated decay due to, 13; albedo effect contributing to problem of, 96–97; dirty snow contributing to, 92–93; Great Lakes–St. Lawrence River system and, 178–79; greenhouse gas (GHG) emissions and, 91–92; impact on water cycle by, 184; impact on water supply by, 171–72; long-term consequences of, 18; as overwhelming concept, 15; scientific evidence confirming, 17. *See also* climate change
Goenka, Debi, 24–25, 32, 33, 39, 40
Goodhope, Leona, 87, 89, 90
Gore, Al, 1
Great Lakes–St. Lawrence River system: efforts to transfer water from, 177; geographic description of, 174; global purchase of water supply from, 174–75; Lake Superior of, 177–78, 182, 189; legal division between Canada and United States, 175; pollution and global warming affecting, 178–79; shipping on the, 179–81. *See also* United States
Great Los Angeles River Cleanup, La Gren Limpieza, 160
Great Pyramid of Giza, 13
The Green Book: The Everyday Guide to Saving the Planet One Simple Step at a Time (Kostigen), 1

Index

green movement, 1–2

greenhouse effect: description and causes of, 91–92; global warming due to, 92

greenhouse gas (GHG) emissions: environmental impact of, 11–12; fossil fuels contributing to, 11; global warming due to, 91–92; Kyoto Protocol attempt to reduce, 69–70; ocean pollution producing, 154

Greenpeace, 30, 44

ground-level ozone, 11

"Hamburger Connection Fuels Amazon Destruction" (CIFOR report), 109

Hammad, Haroot, 8, 9

hazardous waste: amounts and disposal of global, 27–28; Computer Village (Nigeria) landfill site of, 28; dioxin emissions caused by, 95–96; disposal of hospital, 31; exported from the US to India, 26, 27; *Khian Sea* story on illegal disposal of, 130; Mumbai (India) problem with, 27–40. *See also* e-waste; landfills; trash

health problems: air pollution and related, 71, 72; dioxin emissions related to, 95–96; Linfen's environmental-related, 67–68; related to landfills, 140; WHO on air pollution–related deaths and, 106

Hilo trash-art fair, 162

Holt, Adam, 133, 203

Holy of Holies (Jerusalem), 9

Hong Kong People's Council for Sustainable Development, 65

hospital waste disposal, 31

Howell, Colleen, 49, 53, 161

HSBC (Hong Kong and Shanghai Banking Corporation), 47, 48–49

Huascaran National Park (Peru), 14

IGCC (integrated gasification combined cycle), 77

An Inconvenient Truth (film), 1

India: average life span for computer in, 29; Conservation Action Trust of, 53; e-waste problem in, 27–30; hazardous waste exported to, 26–27. *See also* Mumbai (India)

India Travel Guide (Lonely Planet), 34

indigenous people: corporate displacement of Brazilian, 112–13; Inupiat Eskimos, 82–87, 89–90; Orang Asli (Malaysia), 49

Indonesia, 46–47, 110

International Coastal Cleanup Day, 156

Inupiat Eskimos: climate and weather changes noted by, 89–90; cost of relocating the Shishmaref Village, 85–87; traditional lifestyle of, 82–87

Israel: Arava Institute of, 15–16; Society for the Protection of Nature of, 16

Jerusalem: environmental damage occurring to, 5, 7, 9–10; impor-

tance of preserving, 16; religious
and historic significance of, 5;
Western Wall of, 6, 7, 8–10, 11
June gloom, 159
Jury, Bill, 168

Kaimowitz, David, 109
Kamilo Beach, 149, 150–52
Khian Sea story, 130
Knutson, Brian, 178–79
Kyoto Protocol (1997), 69–70

La Niña, 17
Lake Superior, 177–78, 182, 189
Lake Tai contamination
(China), 68
landfills: air pollution from
decomposition of, 131–32;
Computer Village (Nigeria), 28;
declining numbers of available
and open, 127; Deonar landfill
(Mumbai, India), 31–32; difference between dumps and, 137;
Fresh Kills (New York), 123–41;
health hazards associated with
living by, 140; methane gas
emissions of, 132, 138; Puente
Hills landfill (California), 135;
space saved through recycling,
128–29. *See also* e-waste;
hazardous waste; trash; waste
management industry
lead-tainted toys, 67
Lehrer, David, 15–16
Leshan Buddha, 14
Levi "Eco" jeans, 197
Lihong, Wu, 68–69, 74
Linfen (China): coal production and
air pollution of, 61–62, 63–65,

72–74, 79–80; environmental
health problems of, 67–68; high
death rates in, 68; low cost of
living in, 74. *See also* China
logging industry: Borneo logging
camps, 41–42, 45, 55; Brazilian,
110–11. *See also* deforestation
London Telegraph, 118
Lonely Planet's *India Travel
Guide,* 34
Los Angeles Olympics (1984), 66
Los Angeles River basin, 160
Los Angeles Times, 72, 185–86

McDonald's, 108–9
Magellan, Ferdinand, 159
Malaysia: MSC (Multimedia Super
Corridor) of, 50; Orang Asli
people of, 49; rubber industry
of, 110–11. *See also* Borneo
(Southeast Asia)
"Man & Tiger: A Dialogue," 48
Manaus (Brazil), 113–14
Manaus Opera House (Brazil), 111
mangrove forest (Sundarbans Na-
tional Park) [Bangladesh], 53
Marine Pollution Bulletin, 148
Mattel, 37
"Megacities," 22
melting glaciers/ice caps, 97–98
mercury pollution, 164
methane gas emissions, 11, 132, 138
Mississippi River, 156
Mitchell, Andrew, 118–19
Moore, Charles, 145–46, 147, 148,
152, 154, 159–61
Mount Rushmore, 13
MRFs (material recovery facilities),
135–36

MSC (Multimedia Super Corridor)
[Malaysia], 50
Mumbai (India): as case study of
problem of urban coastal populations, 22–25; climate change
consequences faced by, 23–26;
Dharavi slum of, 33–40; flooding problem faced by, 22, 23;
hazardous waste problem in, 27–
40; landslides as threat to, 23;
origins and history of, 25–26;
urban population of, 21–22;
water and sewage deficiencies of,
22–23. *See also* India

National Association of Regulatory
Utility Commissioners, 75
National Oceanic & Atmospheric
Administration (NOAA), 92,
93, 154
National Science Foundation, 97
National Weather Service, 17, 99
Nature Conservancy, 108–9
"navigable waters," 168
Nayokpuk, Herbie "Cannonball," 90
Nayokpuk, Percy, 90
New York Times, 69, 200
9/11, 101, 124
nitrogen oxides emissions: burn-
ing coal producing, 76; EPA
air quality calculation for, 107;
increase of global, 106; as pollu-
tion catalyst, 106
nitrous oxide, 11
Northwestern Hawaiian Islands, 154

Ocean Conservancy, 156
ocean pollution: of Eastern Garbage
Patch, 143–60, 162–64, 196;

environmental importance of
preventing more, 164–65; seabird
deaths due to, 155; seafood mercury buildup from, 164; simple
steps to reduce, 161–62; Western
Garbage Patch (Pacific Ocean),
145. *See also* water supply
The Odd Couple (TV show), 35
Old City (Jerusalem), 8
Olympic Games: Beijing (2008),
66; Los Angeles (1984), 66
Omar, 6
Orang Asli people (Malaysia), 49
orangutans, 50
overpopulation, 10–11

Pacific Ocean: Eastern Garbage
Patch pollution of, 143–60,
162–64; Western Garbage Patch
pollution of, 145
packaging (food), 199–201
palm oil: description and use of,
56; products using, 55–56
palm oil production: Borneo
deforestation due to, 43–45,
46–50, 55–57; changing government policies related to, 45–46;
environmental activism protest-
ing, 44–45; environmental costs
of, 58
particle pollution (particulate
matter), 107
Paschal, Joel, 154–55
Pater, Walter, 205–6
Payne, Junaidi, 45, 56
Petrobras (Brazil), 113
phyotosynthesis of phytophlank-
ton, 153
plankton, 147–48

Plotkin, Mark, 115

polar bears: deformities found recently in, 96; Shishmaref Village stories on, 85

pollution: Arctic, 92–96; ocean, 143–65, 196. *See also* air pollution

POPs (Persistent Organic Pollutants), 95

"prior appropriation" covenant, 176

Puente Hills landfill (California), 135

PVC (polyvinyl chloride), 29

Queens College (New York), 95

Reality Tours and Travel, 34–35, 39

recycling: dirty versus clean MRFs (material recovery facilities), 135–36; dropoff centers and process of, 132–36; EPA's campaign promoting, 126, 127, 128; food packaging, 200–201; landfill space saved through, 128–29; as one simple solution, 100; real environmental benefits of, 139; recycled-content goods versus post-consumer recycled goods, 132; trees saved through paper, 129; in the US, 32–33; wastewater to expand freshwater supply, 185–86. *See also* trash

REDD program (UN), 118–19, 121

Rees, Matt Beynon, 10

Richard, 43, 51, 52, 57–58

Richardson, Bill, 175

Roundtable on Sustainable Palm Oil, 44–45

rubber market (Brazil), 110–11

rubber tree (*Hevea brasiliensis*), 110

Sabah (Malaysia), 46

salt weathering, 13

Sarawak (Malaysia), 46

Saru, 105–6, 116

Schrag, Daniel, 78

seabird deaths, 155

seafood mercury pollution, 164

sea-levels: coastal populations vulnerable to rising, 22–25; melting glaciers and ice caps affecting, 97–98

sewage sludge, 196

Shanxi Museum (Tiayuan, China), 63

Shishmaref Emergency Station (S.E.S.), 84, 90

Shishmaref Village (Alaska): climate change curriculum taught in school of, 98–100; climate changes affecting, 81–98; cost of relocating residents of, 85–86; traditional lifestyle of Inupiat Eskimos of, 82–87. *See also* United States

Silicon Valley Toxics Coalition, 29, 30

Silk Route, 62–63

Society for the Protection of Nature (Israel), 16

soybean production, 108–9

Sri Lanka, 197

Stenek, Ken, 98–100

Stonehenge, 13

Suez Canal, 26

sulfur dioxide: burning coal producing, 12, 76; EPA air quality calculation for, 107

Summit Global Management, 170
Sundarbans National Park (Bangladesh), 53
Sunil, 35, 38, 39
sustainability: Hong Kong People's Council for Sustainable Development roundtable on, 65; WWF-Malaysia approach to Borneo, 45–47, 48–49, 56
"the synthetic sea," 147

Taiyuan (China), 61, 62, 63
Taj Mahal Hotel (Mumbai), 26
tapirs, 116–17
teleconnection patterns: monitoring of ten common, 17; weather significance of, 16
thermal expansion, 184
Third Station of the Cross (Jerusalem), 9
three Rs (reduce, reuse, and recycle), 139
Tocktoo, Stanley, 85, 89, 90, 101
Tower of London, 13
Toxics Link, 27–28
trash: of Eastern Garbage Patch (Pacific Ocean), 143–60, 162–64, 196; EPA definition of, 126; steady increase of produce, 126–27; US generation of, 127; of Western Garbage Patch (Pacific Ocean), 145. *See also* hazardous waste; landfills; recycling; waste management industry
Tsur, Naomi, 16
2008 Watch List (WMF), 13–14

UN Food and Agricultural Organization, 46, 109

UNESCO-designated World Heritage Sites: in China, 14; Sundarbans National Park (Bangladesh) as, 53
United Kingdom, 27
United Nations: Intergovernmental Panel on Climate Change, 77; predicted orangutan extinction by, 50; REDD program of, 118–19, 121
United Nations Environmental Programme, 113
United Nations Framework Convention on Climate Change (1992), 70
United Nations Framework Convention on Climate Change (2007), 69–70
United States: among of e-waste produced in, 33; average life span for computer in, 30; best and worst air quality areas in the, 107; carbon dioxide emissions produced by every person in the, 94; clothing imports and exports from, 197–98; coastal demographics of, 22, 97–98; consumption contributing to global air pollution, 64–67; energy consumption of average home in the, 193; export of hazardous waste produced in, 26, 27; failure to ratify Kyoto pact by, 70; Fresh Kills landfill in the, 123–41; land and agricultural distribution in the, 115–16; nitrogen oxides emissions increase in the, 106; projected state water short-

ages predicted in the, 175–76; vulnerability of coast populations of, 24; waste individually produced and recycled in the, 32–33. *See also* Great Lakes–St. Lawrence River system; Shishmaref Village (Alaska)
University of California, Irvine, 92, 93
upwelling phenomenon, 159
US Army Corps of Engineers, 82, 91
US Department of Agriculture, 108
US Department of Health and Human Services, 148

Vaux, Henry, Jr., 168
Verrazano-Narrows Bridge (New York), 123
Via Dolorosa (Jerusalem), 6, 8, 9, 18
"virtual water," 173–74, 186

Wal-Mart, 65, 200
Wankhade, Kishore, 27, 28, 29, 31, 37
waste management industry, 130. *See also* landfills; trash
water cycle: global warming impact on, 184; process of, 182–83
water pollution: Eastern Garbage Patch (Pacific Ocean), 143–60, 162–64, 196; Great Lakes–St. Lawrence River system and, 178–79; increasing rates of, 167–68; Western Garbage Patch (Pacific Ocean), 145
water shortages: individual actions to prevent, 186–88, 194–96; predicted US state, 175–76;

"prior appropriation" covenant impact on, 176
water supply: annual waste and pollution of, 167–68; declining access to safe, 169–74, 175–76; desalination of, 184; global warming/climate change impact on, 171; Great Lakes, 174–89; individual actions to extend, 186–88, 194–96; Lake Tai contamination (China) of drinking, 68; Mumbai (India) lack of drinking, 22–23; "prior appropriation" of, 176; protection of "navigable waters," 168; recycling wastewater to expand, 185–86; "virtual water," 173–74, 186. *See also* ocean pollution
water vapor, 11
Weather Channel, 156
weather versus climate, 18
West Coast National Park (South Africa), 14
Western Garbage Patch (Pacific Ocean), 145. *See also* Eastern Garbage Patch (Pacific Ocean)
Western Wall (Jerusalem): decaying evidenced in, 6, 7, 9–10, 11, 12; limestone structure of, 12; rock connecting three religions by the, 18; underground tour of excavation of, 8–9; walking along Western Wall Plaza, 9–10
Weyiouanna, Fannie, 87
Weyiouanna, Tony, Sr., 86, 89, 90, 100
Wickham, Henry Alexander, 111
Wilson, E. O., 58
Wired magazine, 66

WMF (World Monument Fund): approach to Borneo sustainability by, 45–47, 48–49, 56; on shared human history importance, 15; 2008 Watch List of, 13–14

World Bank: on deforestation of Indonesian Borneo, 43, 46; global air pollution measurements taken by, 107; on Mumbai sewer system deficiencies, 23

World Health Organization, 56, 106

World Water Council, 173, 174

World Wildlife Fund, 155

Worldwatch Institute, 109

Worn Again shoes, 196–97

Yip, Plato, 65

Zender, Charlie, 93

zooplankton, 147–48, 153

About the Author

PHOTO: JAY LAWRENCE GOLDMAN

Thomas M. Kostigen is coauthor of the *New York Times* bestseller *The Green Book*. He writes the "Ethics Monitor" column for Dow Jones *MarketWatch* and the *Better Planet* column and blog for *Discover* magazine. He is a longtime journalist and former *Bloomberg News* editor. Kostigen has been writing about global warming, the environment, social issues, and government policies for two decades. He lives in Santa Monica, California.